K

BRITISH MEDICAL BULLETIN
VOLUME 55 NUMBER 2 1999

Impact of genomics on healthcare

Scientific Editors

George Poste

John Bell

Kay Davies

Peter Goodfellow

Nick Hastie

Series Editors
L K Borysiewicz PhD FRCP
M J Walport PhD FRCP

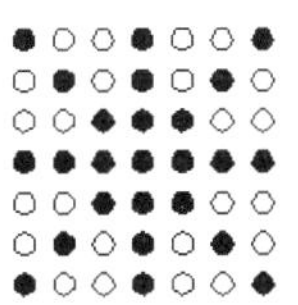

PUBLISHED FOR THE BRITISH COUNCIL BY
THE ROYAL SOCIETY OF MEDICINE PRESS LIMITED

ROYAL SOCIETY OF MEDICINE PRESS LIMITED
1 Wimpole Street, London W1M 8AE, UK
207 E. Westminster Road, Lake Forest, IL 60045, USA

British Library Cataloguing in Publication Data
A catalogue record for this book is available from the British Library
ISBN 1–85315–393–1
ISSN 0007–1420

Subscription information *British Medical Bulletin* is published quarterly in January, April, July and October on behalf of the British Council by the Royal Society of Medicine Press Limited. Subscription rates for Volume 55 (1999) are £145 Europe (including UK), US$250 USA, £149 elsewhere, £73 developing countries. Prices include postage by surface mail within Europe, by air freight and second class post within the USA*, and by various methods of air-speeded delivery to all other countries. Subscription orders and enquiries should be sent to: Publications Subscription Department, Royal Society of Medicine Press Limited, 1 Wimpole Street, London W1M 8AE, UK (Tel +44 (0)171 290 2928; Fax +44(0)171 290 2929; Email rsmjournals@roysocmed.ac.uk).

*Periodicals postage paid at Rahway, NJ. US Postmaster: Send address changes to *British Medical Bulletin*, c/o Mercury Airfreight International Ltd, 365 Blair Road, Avenel, NJ 07001, USA.

Single copies and back numbers of issues published from 1996 are available for purchase directly from the distributors: Hoddle Doyle Meadows Limited, Station Road, Linton, Cambs CB1 6UX, UK (Tel +44 (0)1223 893855; Fax +44 (0)1223 893852). Issues published in 1996 and 1997 cost £24.95/US$41 per copy, and issues published in 1998 and 1999 cost £34.95/US$57 per copy. Please add £2/US$3.50 for postage.

Pre-1996 back numbers: Orders for any title published prior to 1996 should be sent to Jill Kettley, Subscriptions Manager, Harcourt Brace, Foots Cray, Sidcup, Kent DA14 5HP (Tel +44 (0)181 308 5700; Fax +44 (0)181 309 0807).

This journal is indexed, abstracted and/or published online in the following media: Adonis, Biosis, BRS Colleague (full text), Chemical Abstracts, Colleague (Online), Current Contents/ Clinical Medicine, Current Contents/Life Sciences, Elsevier BIOBASE/Current Awareness in Biological Sciences, EMBASE/Excerpta Medica, Index Medicus/Medline, Medical Documentation Service, Reference Update, Research Alert, Science Citation Index, Scisearch, SIIC-Database Argentina, UMI (Microfilms)

Editorial services and typesetting by BA & GM Haddock, Ford, Midlothian, Scotland
Printed in Great Britain by Bell & Bain Ltd, Glasgow, Scotland.

BRITISH MEDICAL BULLETIN Volume 55 Number 2 1999

Impact of genomics on healthcare

Scientific Editors George Poste, John Bell, Kay Davies, Peter Goodfellow and Nick Hastie

Acknowledgements

The planning committee for this issue of the *British Medical Bulletin* was chaired by George Poste and also included John Bell, Kay Davies, Peter Goodfellow and Nick Hastie.

The British Council and the Royal Society of Medicine Press are most grateful to them for their help and advice and particularly for their work as Scientific Editors.

Overview

Heredity is just environment stored.
L. Burbank

The central dogma of modern biology postulates an information flow from DNA to RNA and thence to proteins. The information needed to construct an organism resides in the DNA and the precise sequence of its constituent four bases. The international effort to sequence the human genome will be finished within the next 5 years. Knowledge of the order of the human 3 billion (3×10^9) bases will change the way we think about ourselves and the way we study ourselves. In one sense, the sequence represents a description of what it means to be human and the differences in sequence between our genome, and that of other animals, defines our species identity. Similarly, the identities and differences in sequence between members of the human race both bind us together and define our individuality. Two decades ago, the only way to study the genetic material was indirectly through the science of genetics – the study of inheritance. The invention of methods to sequence DNA lead to genomics – the study of DNA sequences. The twin sciences of genomics and genetics have created a new biology focused on the correlation of DNA sequence (genotype) and phenotypic outcome. This volume considers the impact of this new biology on the practice of medicine.

The underpinning technology for sequencing the human genome has not changed since Fred Sanger invented the fundamental chemical reactions in 1977. Improvements have been almost entirely limited to automation and the amplification of scale (industrialisation). The existence of a copy of the human genome sequence, even a rough draft, means that methods for measuring sequence variation are going to be of central importance for future research and exploitation. Nigel Spurr and colleagues describe the platform technologies of the new biology. My own predictions are that technology for *de novo* sequence determination is going to continue to be based on the robust methods of Sanger and for detecting sequence variation, micro-array technology will provide a cheaper approach than mass spectroscopy. The new biology approach emphasises the importance of studies at the whole genome level. This philosophy is instructing several 'post genomics' methods such as micro-arrays for measuring gene expression of all genes simultaneously, and proteomics – the detection of all proteins in a cell or tissue. The importance of genomics, proteomics, glycomics, etc. suggests the likely evolution of the new science of biomics. For the mercenary minded cybersquatter, it is probably worth registering the company name Biomics Ltd and the web site www.biomics.com as soon as possible.

The detection of sequence variation is the starting point for DNA-based diagnostics. Christopher Mathew is a pioneer in this field who has helped to create a new health services profession dedicated to diagnostic support of the NHS. As described by Dr Mathew, DNA diagnosis started little more than a decade ago with the occasional prenatal diagnosis for sickle cell anaemia and thalassaemia. Today, over 12,000 mutations have been described in over 600 genes. For the most common of monogenic genetic diseases, the confirmation of diagnoses, the testing of carriers and prenatal diagnosis is straight forward. For rarer conditions, the situation is different as many of these mutations are expensive to detect using today's technology. This can pose dilemmas for experimental scientist. In my laboratory in Cambridge, we received several requests to help with both post-natal and prenatal diagnosis of the disease campomelic dysplasia. This disease is caused by mutations in *SOX9*, a gene that we had identified and cloned. As this is a dominant lethal disease, the chances of sequential pregnancies being at risk is very, very low (it would require a chimeric gonad); nevertheless, the reassurance offered to a pregnant woman that her fetus does not carry the mutation that afflicted an earlier pregnancy is enormous. To set up the required assays and to run the assays cost tens of thousands of pounds – too much, at the time, for a NHS laboratory. We took the pragmatic approach of always helping any family that had helped us with our research. The challenge going forward is to expand the services to help all that can benefit and to extend the service from the monogenic diseases to the interpretation of the far more complex multigenic diseases such as cancer, diabetes, cardiovascular disease, and various degenerative diseases. The relationship between gene and phenotype is not always simple.

Similar phenotypes can be generated by mutations in different genes and different mutations in the same gene can cause different phenotypes. An example of the latter phenomenon is the finding that Duchenne muscular dystrophy and Becker muscular dystrophy are caused by mutations in the same gene. As described by Jacques Beckmann, the autosomal recessive progressive muscular dystrophies provide a powerful illustration of the former case. These diseases are rare and show overlapping clinical phenotypes. Linkage analysis and then cloning of the disease genes has clarified the nosology. In a practical sense, these studies help afflicted families make reproductive choices. These studies also form the starting point for defining strategies for finding treatments. They are a *sine qua non* should any gene therapy be contemplated. The muscular dystrophies also point to the complexities of epistatic genetic and environmental interactions affecting phenotype. The difference between monogenic disease and complex disease is only a matter of emphasis. Nevertheless, as discussed by William Cookson, identifying the genes involved in complex diseases is proving to be an arduous task and the impact of the new genetics for these diseases is yet to be seen in the clinic.

A doctor confronted with an ill patient has only a limited number of options. One option is to prescribe a drug. Unfortunately, not every patient responds to drug treatment and an unfortunate few may even suffer an adverse reaction. The response to a drug depends in part on genetic variation in the patient. Roland Wolf and Gillian Smith describe the history and practice of pharmacogenetics – the study of individual genetic variation and its contribution to drug response. Klaus Lindpaintner describes the hope of targetting the right drug to the right person.

Since the time that Robert Koch made his famous postulates, the microbial theory of infectious disease has been universally accepted. This emphasis on the disease-causing micro-organism has often overlooked the important role of the genetics of the host. It is probable that the strongest and most acute genetic selection applied by the environment is generated by disease causing microbes. Adrian Hill summarises the current knowledge about the genetics of host resistance to microbial infection. As might be predicted, host resistance is multigenic and the same techniques that are being used to explore complex genetic diseases have been used to identify the genes responsible for susceptibility and resistance. Notable achievements include the identification and cloning of NRAMP1, which confers susceptibility to Leishmania in the mouse. This gene was mapped using crosses between inbred strains of mice and cloned using positional cloning methods. Subsequent experiments have implicated the same gene in susceptibility to pulmonary tuberculosis in man. Other resistance/susceptibility genes have been identified by association studies, for example, sickle cell trait, thalassaemia and G6PD deficiency with malaria resistance.

Microbial genomes have been exerting selection on the human genome since our species began. In the last 50 years, we have been exerting selection on microbial genomes by flooding the environment with antibiotics. Unfortunately, the life cycle of a genome in the microbial world is short, and many micro-organisms have evolved mechanisms for adapting to environments that are hostile and rapidly changing. The result is bacterial resistance to the medically important antibiotics that have sustained our civilisation since the Second World War. If we do not solve the problem of bacterial resistance to therapy then the impact of human genetics and genomics on medicine will be moot. By a fortunate coincidence, the techniques developed to analyse the human genome can be directly applied to other genomes, including bacteria. The complete sequences of the genomes of several pathogens are now available. Christopher Tang and David Holden explain how combining genome sequence data with genetic analysis can identify new targets for both vaccines and drug therapy. We are in a race: we must make new classes of antibiotics before our current armamentarium fails altogether.

Past eugenic abuses have left a legacy that must be faced by human genetics. This legacy and a general public that is understandable apprehensive can make the presentation of genetic advice difficult – especially in the clinic. Theresa Marteau shares her insights and experiences of communicating complex genetic issues in clinical practice.

We are fortunate to live in a society that allows the discussion of choice. The greatest scientific advances of this century have been generated by quantum physics and biology. The practical exploitation of quantum physics was atomic fission: the nuclear reactor and the atomic bomb. The contingencies of war meant that the construction of the bomb was not debated and the decision to bomb Hiroshima was not discussed in the media. The ethical issues posed by genomics and genetics are not new, but they are acute. It is also true that the science of genetics has been abused in the past. Sandy Thomas discusses the ethical issues surrounding the new biology. As the Director of the Nuffield Council on Bioethics, Dr Thomas has direct experience of the social debate and has heard the arguments from many perspectives. The advances that will allow identification of individuals at risk for disease can be portrayed as being responsible for social stigmatisation of the same individuals. The dangers are real but I believe the benefits are greater – we must argue the case for genetics and genomics, the new biology and the new medicine.

As geneticists and genomicists, the editors of this issue may be guilty of 'genocentricity' forgetting that advances are occurring rapidly in all areas of biology. Peter Morris, a pioneer of organ transplantation, points out that this discipline is also challenged by advances that offer the promise of patient benefit but pose safety and ethical questions.

The benefits of the new medicine will include a better definition and classification of disease, new diagnostic methodologies, identification of factors that predispose to disease and improved therapies tailored to individual patients. If all the hoped-for advances of the new medicine were available today – the impact would not be high. Delivery of health services requires the political desire to invest. The starting points are education of the public and training of the medical profession. Robin Fears, David Weatherall and George Poste present a manifesto for investing in the future of medicine and the British NHS.

Peter N Goodfellow
Senior Vice President

Discovery
SmithKline Beecham Pharmaceuticals
Frontier Science Park
Third Avenue, Harlow
Essex CM19 5AW, UK

New technologies and DNA resources for high throughput biology

Nigel Spurr, Ariel Darvasi, Jonathan Terrett and **Liz Jazwinska**

Biotechnology and Genetics, SmithKline Beecham, Harlow, Essex, UK

The rapid increase in DNA sequencing information is opening up new opportunities in genetics. The current methods for processing and analysing genetic data are, however, slow and labour intensive. The next wave of genetic analysis will rely on the analysis of DNA variation from large population based cohorts. These studies will provide important new data on population and disease genetics and have the potential to make a significant impact on our current healthcare practices. In order for these studies to deliver, we need to develop a new generation of ultra-rapid DNA technologies which will allow us to generate, capture and efficiently exploit these new data. This chapter describes the recent advances in DNA sequencing and genotyping technologies that will lead to 100–1000-fold increases in our ability to produce the DNA data we need to explore and exploit the new genetic opportunities to the full.

The Human Genome Project is poised to deliver the sequence of the human genome to the scientific community. Current estimates are that the first draft of the human genome sequence will be available by mid 2000, and the completed sequence will be available by the end of 2003. This important event will mean that, for the first time, it should be possible to identify all the coding sequences of the predicted 100 000 genes. The genetic analysis of other mammalian genomes (*e.g.* mouse, rat and dog) will not lag far behind and, together with the sequence of microbial genomes, will fuel significant advances in human medicine.

The completed sequence of the human genome followed by the other mammalian genomes, will see an end to many of the current gene by gene mapping projects which formed the foundation for the larger scale full genome sequencing projects. For example, projects using whole genome radiation hybrid mapping panels (*e.g.* human, rat and mouse[1–3]) have made an invaluable contribution to efforts to identify genes leading to disease. These experiments are described in two pivotal publications describing the location of over 30 000 human genes[4,5]. Indeed, it was these original efforts in gene mapping which gave the first indication of the benefits of whole genome analysis. The release of the full human genome sequence will lead to unparalleled efforts to conduct experiments at the

Correspondence to: Dr Nigel Spurr, Biotechnology and Genetics, SmithKline Beecham, New Frontiers Science Park (North), Third Avenue, Harlow, Essex CM19 5AW, UK

British Medical Bulletin 1999;**55** (No. 2): 309–324

whole genome scale as opposed to the current single gene approach; this has important implications for the analysis of human health and disease. An axiom much quoted by many traditional biochemists is '*one gene, one postdoc, one career*', this is now set to change as it will be possible, in one experiment, to analyse all the genes in a single pathway or see the results of a gene knockout on gene expression in specific tissues. We will be equipped to analyse complex genetic questions, which were not feasible with this approach by 'single gene' analysis. For example, by being able to analyse genetic pathways and interactions, we should be equipped to analyse the consequences of drug administration and this will lead to a greater understanding of the safety and efficacy of therapeutics.

In addition, delivery of the complete human genome sequence will advance considerably the science of genetics based on the study of variation. It has been estimated that comparison of the genomes of two individuals would yield up to 3 million differences at the DNA level, approximately one base change every kilobase of sequence. To date, it has only been possible to analyse a small proportion of these differences. If the technology were available to identify rapidly and test for these differences, the field of predictive medicine linking genotype with phenotype could advance significantly. This would open up the field of pharmacogenetics, in particular, our ability to link individual response to drugs with differences seen at the DNA level in specific genes or in regulators of gene expression.

The delivery of so much new information will introduce many new exciting opportunities and challenges: however, do we now have, or will we be able to develop, the necessary technological resources to meet the challenges? This chapter will focus on three areas of emerging technology. These are firstly, improvements to DNA sequencing and, in particular, its application to whole genome sequencing and the analysis of expressed sequence tags (ESTs). Secondly, the new developments in genotyping technology designed to cope with the new class of markers of DNA variation based on single nucleotide polymorphisms (SNPs). Finally, we will address the design and use of microarrays for the analysis of gene expression and specifically the use of ESTs in this rapidly developing area. All of these areas of technology change will lead to a deluge of biological information from the massively parallel technologies that will arrive in the next 5–10 years.

Expressed sequence tags (ESTs)

In addition to whole genome sequencing, there has been a considerable effort over the past 5 years to identify, specifically, the genes expressed in a particular tissue. Coding sequences represent approximately 10% of a mammalian genome and of the estimated 100 000 genes, only about 20% are expressed specifically in any tissue. Therefore, biologists have been interested in obtaining the information on which genes are expressed in any particular tissue. To this end, cDNA libraries have been prepared from mRNA from tissues of interest and random shotgun sequencing of these expressed genes has taken place. Clones in these libraries have been sequenced at random and wherever possible single sequence reads of 100–500 bp or more, have been obtained from both ends of the same clone: these sequences became known as ESTs.

There are now over 1.5 million such sequences in the public domain http://www.ncbi.nlm.nih.gov/dbEST/index.html[6]. To reduce the requirement for re-sequencing the most abundant transcripts in any tissue (mRNA abundance may vary between 1 to 10 000 copies per cell, or more), most analyses have been carried out on normalised cDNA libraries[7,8]. By this process, the use of a pre-hybridisation step substantially reduces the copy number of abundant transcripts.

ESTs can provide information that helps to describe the cDNA in biological terms. For example, a major advantage in sequencing cDNAs from tissues is that information is obtained on the range of genes expressed in a tissue. This type of tissue profiling is important for biologists comparing gene expression in normal and diseased tissue. Additionally, analysis of an EST database can assist in the identification of a specific gene and, in many cases, previously unrecognised members of gene families or even new gene families. It is also possible to cluster together overlapping ESTs to identify a complete cDNA clone for a gene.

There is a considerable effort currently underway to identify a non-redundant set of human and mouse genes (http://www.ncbi.nlm.nih.gov/UniGene/index.html)[9]. Careful bio-informatic analysis of the EST data to date has led to the identification of 30–40 000 human genes, approximately half of the total number of expected genes. The significance of the EST sequencing projects is described later in relation to the use of microarrays as a tool for analysing gene expression.

Genotyping and SNPs

Genetics and the study of inherited variation are inextricably linked. Since 1980, DNA sequence variation has become the hard currency of modern genetics. The first variants detected were differences in fragment length detected using restriction enzymes (RFLPs)[10]. This was rapidly followed by the discovery of variable number tandem repeats (VNTRs) or mini-satellites that form the backbone of our current forensic approach to DNA fingerprinting[11]. RFLPs and VNTRs formed the core of the first whole human genetic linkage maps[12]. In the late 1980s, a new class of variable genome marker was found: these were the tandem nucleotide repeats[13,14]. These repeats appear to be randomly distributed within the human genome and the total length of the tandem repeat is highly variable between individuals. As such, these markers are often multi-allelic and this important characteristic made possible the generation of the whole human genome genetic linkage maps published in the 1990s[15,16].

The majority of highly variable markers occur outside of the coding region of genes. There are several notable exceptions, for example an RFLP occurring within the coding region of the β-globin gene has enabled the identification of individuals susceptible to sickle cell anaemia. This important discovery facilitated the development of a molecular diagnostic test for this disorder and its successful application for many years[17].

In recent years, a new form of DNA marker has come to prominence – this is the single nucleotide polymorphism (SNP). SNPs are the most abundant DNA marker present in the human genome; it is estimated that over 3 million useful SNPs exist in the human genome. Only a few thousand of these, however, have currently been identified. Many SNPs are found in the coding regions of genes and have been termed cSNPs. SNPs within the coding region of a gene can have a range of effects:

- Neutral alteration, base substitution with no effect on the amino acid
- Conservative change, a base substitution with an altered amino acid, but with minimal effect of gene structure or function
- Non-conservative alteration leading to an amino acid change with dramatic effects on gene structure or function. Included in this group are base changes leading to mutations for example, frame-shifts leading to termination or other nonsense events.

It has been proposed that SNPs will be the most important genetic tool for the future, as it will be our ability to detect association between SNPs and the hereditary phenotype that will drive a new revolution in medicine in the next 10 years.

The complete sequence of the human genome will open up a whole new era in population genetics. For the first time, it will be possible to assay across the whole genome to identify variation and it will be possible to design powerful analyses to haplotype in specific regions to measure accurately linkage disequilibrium. This greater potential spectrum for genetic analysis will allow more directed pharmacogenetic/pharmacogenomic studies and potentially lead to a more complex understanding of phenotype and genotype relationships. Lander and Schork reviewed the use of SNPs in the analysis of polygenic diseases and suggested that approximately 100 000 SNPs are required for the conduct of whole genome population association studies[18]. In addition, two other elements are essential for this work. The first is access to well-characterised patient and control populations and the second is access to technologies permitting rapid and accurate genotyping of multiple individuals with numerous markers. For example, a genotyping experiment on 5000 individuals using 100 000 single nucleotide polymorphic markers would generate 500 million genotypes. At present, even the most productive genotyping laboratories can only produce up to 5 million genotypes in a year. Therefore, there is a critical need for more efficient genotyping methods that reduce costs and reagent usage as well as increasing throughput by several logs.

New technologies

Table 1 summarises the new technologies and their most common applications. This is a guidance and not a definitive listing.

Table 1 Common genetic analysis methods for emerging technologies

Method	DNA sequencing	SNP detection	VNTR markers	Gene expression profiling
MALDI-TOF	++	+++	+++	+
CMST (MS)	++	+++	+++	+
Capillary electrophoresis	+++	+++	+++	NA
Silicon chips	+	++	NA	+++
Glass chips	+	+	NA	+++

MS, mass spectroscopy; NA, not applicable; +, least applicable; +++ most applicable.

DNA sequencing

For the past 10 years, DNA sequencing has been dominated by the use of gel-based fluorescent methods which incorporate automated laser detection to capture the fluorescent signal from a range of fluorchromes. All of the commercially available systems use polyacrylamide gels in a plate format with a capacity of 36–96 samples per gel run. Continuous small changes to the sequencing chemistries and improvements to the preparation of the starting template DNA over this period have increased the number of bases read in a single run from around 200 per gel track to over 1000. This fluorescent based sequencing technology has been used to analyse a wide range of viral, bacterial and, more recently, eukaryotic genomes. The recent complete sequence analysis of the 97 Mb genome of the nematode worm *Caenorhabditis elegans* was a substantial sequencing feat amply illustrating the current power of this technology (The *C. elegans* Sequencing Consortium, 1998). However, if the entire human genome is to be completely sequenced by the year 2003, as is the current goal, then further significant changes in technology are still required.

Capillary electrophoresis

There are several reasons why DNA sequencing technology has to advance in the next few years. Any gel-based method is by its very nature slow and cumbersome: in particular, it takes time to prepare each gel and even with a factory style set-up as exemplified by the Sanger centre, and other major centres involved in large scale sequencing projects, this is a bottleneck. An additional problem occurs when using a gel loaded with multiple samples. Using current gel-based automated sequencing systems, failure to generate a sequence from a sample gives rise to a blank track, can cause the automated scanner to misread the position of the gel lanes such that the generated sequences are incorrectly assigned to the parent samples. These difficulties in gel reading can be overcome by using sophisticated computer software to track and mange the data generated. However, it is not simple to generate software able to iron out the faults in a poor quality gel.

To overcome many of the difficulties which are currently limiting the sequencing throughput using gel based technologies, several companies and academic groups have decided to look for alternative sequencing technologies in an effort to reduce labour costs, save time and increase accuracy. One such solution has been the development of DNA sequencers that use bundles of fine capillaries. The basic principle of this method – capillary electrophoresis using a single capillary scanned by a laser – has been used for several years both in DNA and protein chemistries. The unique feature of the new generation of instruments is that the throughput capacity has been increased approximately 100-fold through the use of capillary bundles comprising 96 capillaries, and rapid and accurate scanning using confocal microscopy. Fundamentally, the same DNA sequencing chemistry can be employed in these machines as is used in gel-based systems, although the increased sensitivity of the capillary method leads to a decrease in volumes and concentrations of the reagents and DNA template. A key advantage of the capillary machines is that each sample is placed in a single capillary, if a reaction fails then that sample will be blank and no data will be collected. In this way, the problem of mis-reading samples that can occur in a gel-based system, is eliminated. A further advantage of the capillary system is the reduced set-up and throughput times. A standard slab gel systems takes approximately 4–5 h to prepare and run while a capillary system can run 96 samples in under 2 h. Thus, the capacity and throughput of the capillary machines is around 4 times greater that of a standard DNA sequencer and a single capillary sequencer can produce over 1 Mb of DNA sequence in an average week.

Genotyping: mutation detection and gene expression technologies

Several methods have been developed to detect single nucleotide variation in genomic DNA. Some of the methods – such as the direct sequencing of DNA, denaturing HPLC or single strand conformation polymorphism analysis – are well suited to detecting variants, but are only applicable for limited numbers of samples or pools of samples. Other methods exist for the parallel detection of variants in large numbers of independent samples.

Restriction site analysis

In the 1980s, the first gene mutations were detected using alterations in restriction enzyme recognition sites (restriction fragment length polymorphisms or RFLPs). This methodology was of limited application as it was restricted by the frequency of suitable enzyme cutting sites. However, it is still possible to design such assays and they remain a popular method for detecting variants when combined with PCR amplification of specific regions of DNA (amplified restriction fragment length polymorphism or ARFLPs) rather than the restriction enzyme digestion of whole genomic DNA. By the method of ARFLP analysis, both alleles can be identified in one reaction and, although considered to be a reasonably satisfactory method, the overall costs are high (Taq DNA polymerase and a suitable restriction endonuclease enzyme are required) and it is both labour intensive and time consuming in its set up and analysis.

Allele specific oligonucleotides

Although many different technologies are now being developed for automated SNP assays, one of the easiest to use is amplification refractory mutations (ARMs), developed by Zeneca in 1988. ARMs, also known as allele specific amplification (ASA), assays rely on the ability (or not) of Taq polymerase to extend a primer bound to DNA template where the 3′ base of the primer matches its complementary base in the template. This generates ± scores for each allele, so that two PCRs are required for each DNA sample. Single tube assays can be developed using ASA primers of different length or with different 5′ fluorophores. In these cases, the PCR products are analysed on gels. A number of electrophoresis formats are available: e.g.

- Agarose gels: limited sample capacity, low cost for equipment and reagents
- Acrylamide slab gel systems: large sample capacity, high costs for equipment and reagents
- MADGE: agarose or acrylamide in a microtitre plate format – limited sample capacity but automated loading allowing high throughput at low cost)[20]
- Capillary matrix separations: large capacity, rapid and accurate (see DNA sequencing section).

Some fluorescent technologies can eliminate the need for gel analysis, *e.g.* TaqMan[21] fluorescence polarisation[22] and free phosphate accumulation[23]. Other types of fluorescence-based genotyping techniques not requiring gel analysis often require extra steps, *e.g.* fluorescence resonance energy transfer (FRET)[24], which uses a post PCR oligonucleotide ligation assay (OLA), mini-sequencing[25] and dynamic allele-specific hybridisation[26]. An additional genotyping technique which is PCR independent and shows great promise is the homogeneous isothermal technique which involves using a cleavase enzyme[27] Two oligonucleotides are needed for this method, one of the oligonucleotides recognises the SNP and overlaps the other oligonucleotide by at least one base. The second oligonucleotide has a 5′ overhang of bases that are not complementary to any DNA being assayed. These two oligonucleotide primers form a 3-way flap junction that is recognised by the cleavase enzyme only when the SNP oligonucleotide matches the allele being assayed. The cleaved fragment can be visualised on gels or by fluorescence. Currently, two assays are required for each genotype.

At present, one of the most widely used methods for rapid genotyping employs allele specific oligonucleotides arranged in a dot blot format, often on a nylon membrane, and labelled PCR products are hybridised against the allele specific oligonucleotides. This method is commonly used for HLA and cystic fibrosis genotyping where rapid and accurate analysis of multiple polymorphisms are required. The very close proximity of the polymorphisms in these systems makes this 'reverse dot-blot' more efficient than single polymorphism PCR assays as a single PCR product can contain multiple polymorphisms[28]. The goal of all new genotyping developments is to reduce cost to a minimum and, where appropriate, to maximise throughput. These two goals are conditions that have to be met if we are to analyse the genetic factors underpinning complex traits such as response to drug treatment or predisposition to diabetes and schizophrenia.

Mass spectroscopy

A major area of emerging technology is the use of the mass spectrometer in the analysis of DNA. This has the advantage of eliminating the need for gel-based analysis of DNA fragments used in both DNA sequencing and genotyping methods. Two types of mass spectroscopy are currently being utilised. Matrix-assisted laser desorption ionisation time of flight spectrometry (MALDI-TOF) and cleaveable mass spectrometry tags (CMSTs). The first of these methods measures the precise mass of small DNA fragments that have been ionised and accelerated through a high vacuum. It is precise enough to measure the differences between two nucleotide bases. The smallest mass difference being between adenine and thymidine a total of 9 Da. In MALDI-TOF the mass of a specific DNA fragment is measured. In CMST spectroscopy, small molecules are photochemically cleaved from oligonucleotides and these are then fired down the mass spectrometer tube.

For DNA genotyping, both methods use standard PCR amplification of a region containing a polymorphisms. The techniques can be used to detect both single nucleotide changes as well as to assay for di-nucleotide repeat markers. In MALDI-TOF analysis, a PCR fragment is generated; this double stranded product is heated to generate two strands and one of these is captured using, for example, streptavadin coated magnetic beads and biotin incorporated into one of the primers used for the original PCR amplification. The purified product is then hybridised with a new oligonucleotide primer that binds close to the site of the polymorphism. A primer extension reaction is carried out using dye terminators which adds one or two extra bases to the oligonucleotide primer. The base added will depend on the variant in an individual strand of DNA. The extended primer and the other DNA strand are separated once more and the primer product is fired into the mass spectrometer. The difference in the mass of any extended product can then be accurately measured. In this way, it is possible to detect heterozygotes and homozygotes.

In CMST, a different strategy is followed. The starting point is again a PCR amplified product and the single stranded products are captured onto a solid support, but, in this case, modified oligonucleotides are hybridised to detect the normal and variant sequences. The oligonucleotide primers are modified with photocleavable linkers attached to a small molecule tag of variable composition; this results in the primers having a different final mass. One tag is attached to the oligonucleotide detecting the wild type sequence and the other is specific for the variant sequence. After hybridisation, the excess primers are removed and the double stranded DNA separated and exposed to a short burst of intense light degrading the linker and releasing the mass tag. This product is then fired down the mass spectrometer for analysis. In homozygotes, only a single tag will be detected and in heterozygotes two tags will be seen.

Both MALDI-TOF and CMSTs have been used to genotype SNPs and other markers in individuals[29–33]. In addition, mass spectrometry has been used to analyse the p53 gene[34] and it can be used more generally to carry out small scale DNA sequencing of up to 100 bases[35,36].

There are several problems in the application of mass spectrometry for genome analysis, particularly in DNA sequencing, where only short fragments of less than 100 bases in length can be measured accurately. However, recently, an alternative form of mass spectroscopy using infrared MALDI has been used to analyse fragments of DNA more than 2000 bases in length[37]. This new development opens up the possibility of analysing longer sequence fragments and supports the potential utility of this tool for not only the detection of multiple variants in a single analysis. MALDI-TOF requires an extremely clean product and involves several complex steps prior to the analysis. One group has found that modifying the DNA with alkylation and the incorporation of phosphorothrioate into the DNA backbone reduces the number of steps[38].

Recently, an interesting extension of mass spectroscopy has been used to link the science of genomics with proteomics. Proteomics is the study of proteins and function using massive parallel approaches. Neubauer and colleagues described the analysis by 2D-gel electrophoresis of the multi-protein splicesome complex of a cell. This region is involved in the removal of introns from mRNA precursors and, in this study, the protein components were separated and individual proteins were analysed by nanoelectrospray mass spectroscopy following tryptic digest of the protein. The resulting products were analysed and several runs of amino acid sequence were obtained. The sequence generated was used to search the public EST databases and a number of new genes involved in the complex were identified[39]. This example demonstrates the versatility and power of combining different techniques can result in generating significant amounts of new information to answer complex questions.

DNA chips and microarrays

Since 1996, the two competing technologies of DNA chips and microarrays have become the mainstay of high-resolution genetic analysis. The extraordinary rise of this technology and its importance in the new area of genetic analysis was recently illustrated in the publication of a supplement to *Nature Genetics* entitled the *Chipping Forecast*[40] describing the latest technology developments and their many applications.

Although the two technologies are based on different methods for preparing DNA arrays, they both rely on the hybridisation of complementary strands of DNA, one of which is labelled with a fluorescent tag, to enable the automated detection of hybridising pairs of sequences. The results are collected into a computer database for interpretation. Indeed one of the key features of all such DNA array-based technologies is that access to high quality informatics is essential for the correct interpretation of the data generated.

The first significant technology using DNA chips was developed by the biotechnology company Affymetrix, and uses photolithography as developed in the electronics industry to control the direct synthesis of oligonucleotides onto an etched silicon surface[41]. Oligonucleotides of predetermined sequence of between 15–20 bases can be synthesised in this fashion and up to 65 000 probes can be prepared on a single chip approximately 1.5 cm^2 in size. This elegant technology has many applications in the field of diagnostics and genomics and, in particular, holds great promise in the area of gene expression profiling.

In the case of diagnostics, oligonucleotide DNA chips have been used to detect mutations in the BRCA1 gene in patients with breast cancer, and the complete human mitochondrial genome has been synthesised and used to scan for new mutations[42]. Furthermore, genotyping and species identification of *Mycobacterium* strains has been carried out by arraying the complete genome of *Mycobacterium tuberculosis* and detecting clinically important rifampin resistant strains[43]. Deletion mutants in the yeast *Saccharomyces cerevisiae* have also been detected using this technology[44].

Genomic analyses using DNA chips have been focused around their application in the identification of mutations, the analysis of SNPs and the detection of changes in gene expression. In the first two applications, genomic DNA is fluorescently labelled and hybridised to the arrayed oligonucleotides, whereas for gene expression analysis, mRNA is isolated from a tissue of interest and single strand cDNA is synthesised incorporating a fluorescent label. This enables the comparison of different cDNA samples in the same experiment (for example a normal tissue versus a tumour) using dual colour laser scanning and two fluorescent labels with different excitation wavelengths.

Wang and colleagues described the identification, mapping and genotyping of 3241 new SNPs involving the using of oligonucletotide chips covering 2.3 megabases of genomic DNA[45]. Oligonucleotide arrays have been used to look at expression in a range of human genes and genes of other species[46,47].

DNA oligonucleotide arrays as developed by Affymetrix are relatively expensive and, until recently, access to this technology has been limited. Few academic groups have had an opportunity to work with this technology and to develop it to its full potential. Another issue is the inefficiency of the oligonucleotide synthesis directly onto the silicon surface. This limits the length of the oligonucleotides to around 15–25 bases. An alternative strategy that has been more accessible is outlined below.

At Stanford University, Brown and colleagues have developed a method to attach fragments of DNA to a modified glass surface with polylysine[48] as an alternative to oligonucleotide arrays. In this method, DNA fragments from 50–2000 bases are placed onto a chemically modified surface, for example a glass microscope slide. Another method of chemical attachment is the use of a covalent attachment using silane chemistry. The DNA is attached as a single strand fragment and is then ready for hybridisation with a target molecule comprising either a single gene or a complex of single stranded cDNA prepared from mRNA from a tissue of interest. The target molecules are labelled with a fluorescent probe and, after hybridisation and washing to remove unbound material, the slides are scanned with a fluorescent scanner to detect the position of double stranded signals. The intensity of the signal can be quantified with the amount of molecular target present in the hybridisation solution.

One of the key factors defining the success of microarrays is the choice of DNA fragments for attachment to the slides. These fragments may comprise specific oligonucleotides for the detection of DNA variation or mutation, or can be fragments of a gene for the analysis of expression. The former set of fragments requires that a probe should be at least 50 bases in length to overcome problems encountered in hybridising small molecules in this format. Recently, one biotechnology company has reported that the tethering of the oligonucleotide with either oligonucleotide A or T or chemical linkers allows shorter lengths of sequence to be attached. The latter application of gene expression is gaining significant prominence and is likely to be a major force in genomics laboratories in the future. For gene expression analysis, groups have used ESTs as the probe for complex mixtures of molecular targets. To prepare sufficient material to produce several replicate slides, PCR is used to amplify the EST. The majority of ESTs are cloned in a standard set of vectors and this allows the use of a universal set of oligonucleotide primers to amplify a large number of different genes. These PCR amplified products are cleaned up to remove excess PCR primers and are then ready for spotting onto micro-array grids. A typical grid can contain up to 10 000 different genes in approximately 1 cm^2. The PCR products are arrayed using robotic spotters that can accurately deposit material into a predefined space on the grid. The first study utilising microarrays reported the analysis of 1000 genes in a single experiment[49]. A recent review by Schena and colleagues and others in the *Chipping Forecast* summarises the detail of the methodologies for chip production and highlights the interests of biotechnology companies involved in this field[50]. Other key factors under development are glass attachment chemistries. Several manufacturers of automated robotic instruments are partnering with groups to develop new glass and attachment chemistries that decrease the signal backgrounds and improve robustness and reproducibility.

As discussed above, a considerable effort has been made to generate a set of non-redundant human EST representing many of the expressed human genes. These can be hybridised with cDNA to identify which genes are expressed in a specific tissue and to look for changes in the disease state or to look for alterations in expression following cell stimulation with an exogenous compound. For example, several groups have analysed the expression of genes in normal and tumour tissue in breast cancer and in other tumours including rhabdomyosarcoma and Ewing's tumour[51–53]. Cell lines have been analysed for the effect of stimulating quiescent fibroblast cells following the addition of serum[54].

Another major area of application has been to use microarrays to carry out whole genome analysis in a single experiment. The completion of the yeast *S. cerevisiae* genome[55] has led to several studies in this area. In one such study, the 2479 yeast open reading frames were PCR amplified, arrayed and hybridised with a range of cDNAs to understand the biology of the yeast genome[46]. Other studies have analysed the regulation of glucose metabolism and the transcriptional control of sporulation[56,57] and a similar set of studies have been conducted using the ESTs generated from the genetic analysis of the plant genome of *Arabidopsis thaliana*. In this case, 1400 ESTs were arrayed and differential expression in the major components of the plant were analysed[58].

In the pharmaceutical industry, similar experiments are underway looking at the effect of therapeutics on cell lines and tissues from animals and patients. This analysis is being used to help identify the role of new drug targets and also to understand the secondary effects of these targets in pathway analysis. Similar work is being carried out to analyse the toxicity and safety of new therapeutics by analysing the profiles of expression of known mutagens, teratogens and other tissue toxic compounds[59,60].

References

1 Cox DR, Burmeister M, Price ER, Kim S, Myers RM. Radiation hybrid mapping: a somatic cell genetic method for constructing high-resolution maps of mammalian chromosomes. *Science* 1990; **250**: 245–50

2 McCarthy LC, Terrett J, Davis ME *et al.* A first generation whole genome-radiation hybrid map spanning the mouse genome. *Genome Res* 1997; **7**: 1153–61

3 Walter MA, Spillett DJ, Weissenbach J, Goodfellow PN. A method for constructing radiation hybrid maps of whole genomes. *Nat Genet* 1994; **7**: 22-8

4 Deloukas P, Schuler GD, Gyapay G *et al.* A physical map of 30,000 human genes. *Science* 1998; **282**: 744–6

5 Schuler GD Boguski MS, Stewart EA *et al.* A gene map of the human genome. *Science* 1996; **274**: 540-6

6 Bouguski MS, Lowe TM, Tolstoshev CM. dbEST-database for "expressed sequence tags". *Nat Genet* 1993; **4**: 332–3

7 Soares MB, Bonaldo MF, Jelene P, Su S, Lawton L, Efstratiadis A. Construction and characterization of a normalised cDNA library. *Proc Natl Acad Sci USA* 1994; **91**: 9228–32

8 Soares MB. Identification and cloning of differentially expressed genes. *Curr Opin Biotechnol* 1997; **8**: 542–6

9 Schuler GD. *J Mol Med* 1997; **75**: 694–8

10 Botstein D, White RL, Skolnick M, Davis RW. Construction of a genetic linkage map in man using restriction fragment length polymorphisims. *Am J Hum Genet* 1980; **32**: 314–31

11 Gill P, Jeffreys AJ, Werrett DJ. Forensic application of DNA fingerprints. *Nature* 1985; **318**: 577–9

12 Donis-Keller H, Green P, Helms C *et al.* A genetic linkage map of the human genome. *Cell* 1987; **51**: 319–37

13 Litt M, Luty JA. A hybervariable microsatellite revealed by in-vitro amplification of a dinucleotide repeat within the cardiac muscle actin gene. *Am J Hum Genet* 1989; **44**: 397–401

14 Weber JL, May PE. Abundant class of human DNA polymorphisims which can be typed using the polymerase chain reaction. *Am J Hum Genet* 1989; **44**: 388–96

15 Dib C, Faure S, Fizames C *et al.* A comphrehensive genetic map of the human genome based on 5264 microsatellites. *Nature* 1996; **380**: 152–4

16 Gyapay G, Morissette J, Vignal A *et al.* The 1993-94 Genethon human genetic linkage map. *Nat Genet* 1994; **7 (2 Spec)**: 246–339

17 Kan YW, Dozy AM. Antenatal diagnosis of sickle-cell anaemia by DNA analysis of amniotic-fluid cells. *Lancet* 1978: ii: 910–2

18 Lander ES, Schork NJ. Genetic dissection of complex traits. *Science* 1994; **265**: 2037–48

19 The *C. elegans* Sequencing Consortium. Genome Sequence of the Nematode C. elegans: a platform for investigating biology. *Science* 1998; **282**: 2012–8

20 Day IN, Spanakis E, Palamand D, Weavind GP, O'Dell SD. Microplate-array diagonal-gel electrophoresis (MADGE) and melt-MADGE: tools for molecular-genetic epidemiology. *Trends Biotechnol* 1998; **16**: 287–90

21 Whitcombe D, Brownie J, Gillard HL et al. A homogeneous fluorescence assay for PCR amplicons: its application to real-time, single-tube genotyping. *Clin Chem* 1998; **44**: 918–23

22 Gibson NJ, Gillard HL, Whitcombe D, Ferrie RM, Newton CR, Little S. A homogeneous method for genotyping with fluorescence polarization. *Clin Chem* 1997; **8**: 1336–41

23 Gibson NJ, Newton CR, Little S. A colorimeteric assay for phosphate to measure amplicon accumulation in polymerase chain reaction. *Anal Biochem* 1997; **254**: 18–22

24 Chen X, Livak KJ, Kwok PY. A homogeneous, ligase-mediated DNA diagnostic test. Genome Res 1998; **8**: 549–56

25 Pastinen T, Kurg A, Metsplau A, Peltonen L, Syvanen AC. Minisequencing: a specific tool for DNA analysis and diagnostics on oligonucleotide arrays. *Genome Res* 1997; **7**: 606–14

26 Howell WM, Jobs M, Gyllensten U, Brookes AJ. Dynamic allele-specific hybridization. A new method for scoring single nucleotide polymorphisms. *Nat Biotechnol* 1999; **17**: 87–8

27 Marshall DJ, Heisler LM, Lyamichev V *et al.* Determination of hepatitis C virus genotypes in the United States by cleavase fragment length polymorphism analysis. *J Clin Microbiol* 1997; **35**: 3156–62
28 Serre JL, Taillandier A, Mornet E, Simon-Buoy B, Boue J, Boue A. Nearly 80% of cystic fibrosis heterozygotes and 64% of couples at risk may be detected through a unique screening of four mutations by ASO reverse dot blot. *Genomics* 1991; **11**: 1149–51
29 Braun A, Little DP, Koster H. Detecting CFTR gene mutations using primer oligo base extension and mass spectrometry. *Clin Chem* 1997; **43**: 1151–8
30 Haff LA, Smirnov IP. Multiplex genotyping of PCR products with MassTag-labeled primers. *Nucleic Acids Res* 1997; **25**:3749–50
31 Haff LA, Smirnov IP. Single-nucleotide polymorphism identification assays using a thermostable DNA polymerase and delayed extraction MALDI-TOF mass spectrometry. *Genome Res* 1997; **7**: 378–88
32 Laken SJ, Jackson PE, Kinzler KW *et al.* Genotyping by mass spectrometric analysis of short DNA fragments. *Nat Biotechnol* 1998; **16**: 1352–6
33 Ross P, Hall L, Smirnov I, Haff L. High level multiplex genotyping by MALDI-TOF mass spectrometry. *Nat.Biotechnol* 1998; **16**: 1347–51
34 Fu DJ, Tang K, Braun A *et al.* Sequencing exons 5 to 8 of the *p53* gene by MALDI-TOF mass spectrometry. *Nature Biotechnol* 1998; **16**: 381–4
35 Fu DJ, Broude NE, Koster H, Smith CL, Cantor CR. A DNA sequencing strategy that requires only five bases of known terminal sequence for priming. *Proc Natl Acad Sci.USA* 1995; **92**: 10162–6
36 Koster H, Tang K, Fu DJ *et al.* A strategy for rapid and efficient DNA sequencing by mass spectrometry. *Nature Biotechnol* 1996; **14**: 1123–8
37 Berkenkamp S, Kirpekar F, Hillenkamp F. Infrared MALDI mass spectrometry of large nucleic acids. *Science* 1998; **281**: 260–2
38 Gut IG, Jeffery WA, Pappin DJC, Beck S. Analysis of DNA by 'charge tagging' and matrix-assisted laser desorption/ionization mass spectrometry. *Rapid Commun Mass Spectrom* 1997; **11**: 43–50
39 Neubauer G, King A, Rappsilber J *et al.* Mass spectrometry and EST-database searching allows characterization of the multi-protein spliceosome complex. *Nature Genet* 1998; **20**: 46–50
40 Phimister B. The Chipping Forecast. Supplement (ed). *Nature Genet* 1999; **21 (Suppl)**: 1–62
41 Lipshutz RL, Fodor SPA, Thomas R, Gineras TR, Lockhart DY. High density synthetic oligonucleotide arrays. *Nat Genet* 1999; **21**: 20–4
42 Chee M, Yang R, Hubbell E *et al.* Accessing genetic information with high-density DNA arrays. *Science* 1999; **274**: 610–7
43 Gingeras TR, Ghandour G, Wang E *et al.* Simultaneous genotyping and species identification using hybridization pattern recognition analysis of generic *Mycobacterium* DNA arrays. *Research* 1999; **8**:435–47
44 Shoemaker DD, Lashkari DA, Morris D, Mittmann M, Davis RW. Quantitative phenotypic analysis of yeast deletion mutants using a highly parallel molecular bar-coding strategy. *Nature Genet* 1996; **14**: 450–6
45 Wang DG, Fan J-B, Siao C-S *et al.* Large scale identification, mapping and genotyping of single-nucleotide polymorphisms in the human genome. *Science* 1998; **280**: 1077–82
46 Lashkari DA, DeRisi JL, McCusker JM *et al.* Yeast microarrays for genome wide parallel genetic and gene expression analysis. *Proc Natl Acad Sci USA* 1997; **94**: 13057–62
47 Lockhart DJ, Dong H, Byrne MC *et al.* Expression monitoring by hybridization to high-density oligonucleotide arrays. *Research* 1996; **14**: 1675–80
48 Schena M, Shalon D, Davis RW, Brown PO. Quantative monitoring of gene expression patterns with a complimentary DNA microarray. Science 1995; **270**: 464–70
49 Schena M, Shalon D, Heller R, Chai A, Brown PO, Davis RW. Parallel human genome analysis: microarray-based expression monitoring of 1000 genes. *Proc Natl Acad Sci USA* 1996; **93**: 10614–9
50 Schena M, Heller RA, Theriault TP, Konrad K, Lachenmeier E, Davis RW. Microarrays: biotechnology's discovery platform for functional genomics. *Trends Biotechnol* 1998; **16**: 301–6

51 DeRisi J, Penland L, Brown PO *et al.* Use of a cDNA microarray to analyse gene expression patterns in human cancer. Nature Genet 1996; **14:** 457–60
52 Khan J, Simon R, Bittner M *et al.* Gene expression profiling of alveolar rhabdomyosarcoma with cDNA microarrays. *Cancer Res* 1998; **58:** 5009–13
53 Welford SM, Gregg J, Chen E *et al.* Detection of differentially expressed genes in primary tumour tissues using representational differences analysis coupled to microarray hybridization. *Nucleic Acids Res* 1998; **26**: 3059–65
54 Iyer VR, Eisen MB, Ross DT *et al.* The transcriptional program in the response of human fibroblasts to serum. *Science* 1999; **283:** 83–7
55 Mewes HW, Albermann K, Bahr M *et al.* Overview of the yeast genome. Nature 1997; **387 (Suppl):** 7–8
56 Chu S, DeRisi JL, Eisen M *et al.* The transcriptional program of sporulation in budding yeast. *Science* 1998; **282:** 699–705
57 DeRisi J, Iyer VR, Brown, PO. Exploring the metabolic and genetic control of gene expression on a genomic scale. *Science* 1997; **278:** 680–6
58 Ruan Y, Gilmore J, Conner T. Towards *Arabidopsis* genome analysis: monitoring expression profiles of 1400 genes using cDNA microarrays. *Plant J* 1998; **15**:821–33
59 Debouck C, Goodfellow PN. DNA microcarrays in drug discovery and development. *Nat Genet* 1999; **21 (Suppl):** 48–50
60 Marton MJ, DeRisi JL, Bennett HA *et al.* Drug target validation and identification of secondary drug target effects using DNA microarrays. *Nat Med* 1998; **4**: 1293–1301

DNA diagnostics: goals and challenges

Christopher G Mathew

South Thames (East) Regional Genetics Centre, Guy's and St Thomas's Hospitals Trust, Guy's Hospital, London, UK

The exponential increase in the discovery of human disease genes over the past 10 years has transformed DNA diagnostics from a minor research-based activity to a major professional operation. Mutation testing and linkage analysis are now used to provide prenatal or postnatal diagnosis for a wide range of monogenic disorders, but robust automated procedures for scanning disease genes for mutations are not yet available. The discovery of genes which confer susceptibility to common disorders is likely to create demand for high throughput testing for specific mutations or clinically relevant polymorphisms. Widespread genetic testing must be supported by adequate genetic counselling and by education of healthcare professionals in order to ensure the appropriate application of this information for the benefit of patients and their families.

The provision of DNA diagnostics for genetic disorders is a youthful occupation which is little more than a decade old. What began as a research activity in a few laboratories is now a fully fledged professional activity with its own national organisations, training and qualifications, which provides thousands of genetic tests every day. This transformation was made possible by the extremely rapid pace of disease gene discovery, and by the development of techniques for the detection of specific DNA sequences. First among these was Southern blotting, and later the diagnostic field was transformed by the polymerase chain reaction (PCR) which permits amplification of specific DNA sequences from very small quantities of genomic DNA (see Weatherall[1] for a review of methodology). This review considers what can and is being done in DNA diagnostics, what may be done in the future, and what principles should underpin our use of genetic information.

Correspondence to: Professor CG Mathew, Division of Medical and Molecular Genetics, GKT School of Medicine, 8th Floor Guy's Tower, Guy's Hospital, London SE1 9RT, UK

Human disease genes: a growth industry

Ten years ago, only a handful of genes which caused genetic disorders in humans had been identified. Today, at least 12 000 different mutations have been detected in more than 600 different genes, and this number is growing rapidly[2]. Most of these are associated with Mendelian disorders such as cystic fibrosis, in which a single, rare gene is mutated in a family and confers a very high risk of developing the disease. Although the frequency of each of these disease genes is low in most populations, the cumulative frequency of several thousand such genes adds up to a significant healthcare burden, quite apart from individual tragedy for the affected families. We are also beginning to see other kinds of genes being added to the list, genes which are common in the general population and which confer a moderate degree of genetic susceptibility to common disorders such as diabetes and asthma. Some of these genes have been localised to specific chromosomal regions by typing very large numbers of families with polymorphic DNA markers and demonstrating excess marker sharing in affected siblings[3], and others have been identified by association studies of candidate genes. Homozygosity for the E4 allele of the APOE gene, for example, has been shown to confer increased risk of, and reduced age of onset in Alzheimer's disease[4]. Although the number of such genes which have actually been identified is still small, there is a strong expectation that the number will grow very rapidly in the next 5–10 years.

Mutation detection

Once the disease gene has been identified, a direct diagnostic test requires knowledge of the specific mutation in a particular family. This may not be as easy as it sounds: the cystic fibrosis (CF) gene, for example, has more than 600 different mutations, many of which have been found in only one or a few families, and these mutations are scattered throughout the gene. It may, therefore, be necessary to scan the entire coding sequence of the relevant gene to identify the single nucleotide which is altered. Obviously this can be done by sequencing the gene, but even with the help of automated DNA sequencers this is labour-intensive and expensive, particularly if the gene in question happens to be large. Many disease genes have coding sequences which are 5000–6000 base pairs (bp) in length. A further complication is that most genes are highly interrupted, with the coding sequence often divided into as many as 40–50 exons. These must either be amplified individually from genomic DNA, or the mRNA can be reverse transcribed and amplified by PCR in one or a few large fragments for subsequent analysis. A variety of mutation scanning methods have been developed[5], such as chemical cleavage, in which a heteroduplex DNA molecule formed between the mutant and a wild-type DNA is chemically modified and cleaved as a result of distortion of the double

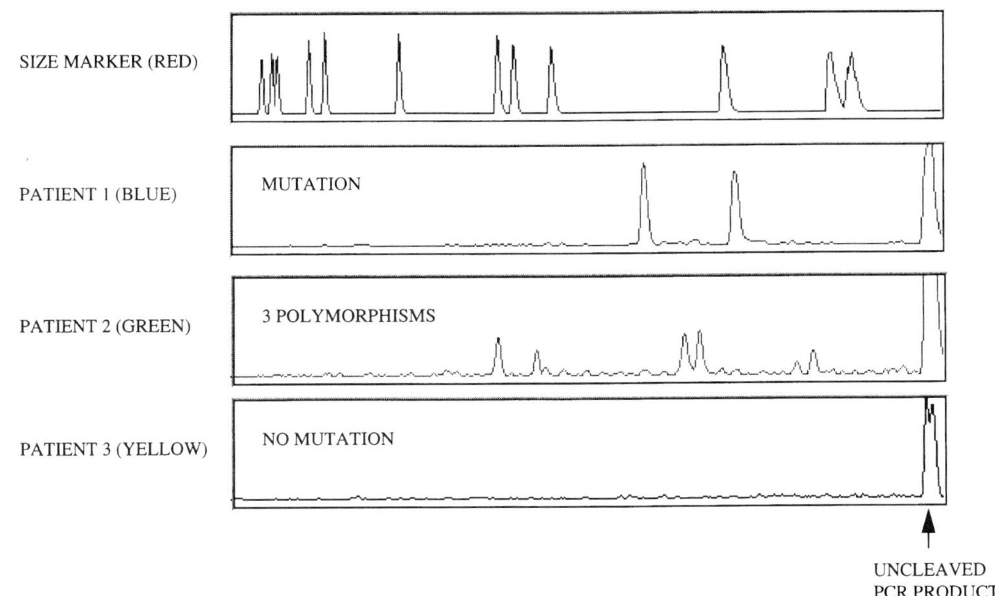

Fig. 1 Mutation scanning of exon 11 of the *BRCA1* gene by chemical cleavage analysis in three patients. The PCR product from each patient has been labelled with a different fluorescent dye, allowing analysis of all three patients on one lane of a fluorescent DNA analyser. In the sample from patient 1 the full length PCR product of 1519 bp has been cleaved into 2 fragments of 843 and 676 bp – DNA sequencing of the region of mismatch revealed a deletion of a C nucleotide at position 2360 of the coding sequence of the *BRCA1* gene. Three common polymorphisms were detected in patient 2, and no mutations were found in patient 3.

helix at the position of the sequence mismatch, and the cleaved fragments are then detected and sized by gel electrophoresis[6]. The mutation must be characterised by sequencing the region of cleavage. Several different fragments can be analysed simultaneously if they are labelled with fluorescent dyes of different colours and separated on a fluorescent DNA analyser, thus permitting scanning of as much as 4000 bp for a mutation in one lane of the gel with high sensitivity[7]. The use of this technique to detect a mutation in the *BRCA1* gene, which causes predisposition to breast and ovarian cancer, is shown in Figure 1.

Once the causative mutation has been identified, a simple PCR assay can be devised to test for it in other family members. In the case of mutations of a single nucleotide, diagnosis is usually by use of a restriction enzyme which has an altered fragmentation pattern at the site of the mutation, or by use of specific PCR primers which will only amplify from the normal or from the mutant allele. Mutation detection in disorders which result from expanded triplet repeat sequences[8], such as Huntington's disease, is relatively straightforward since the region of the gene which undergoes expansion can be amplified by PCR, and the products sized by gel electrophoresis. PCR can also be used to detect the large deletions which occur within the dystrophin gene in Duchenne muscular dystrophy[9], and, with the help of fluorescent DNA analysers, to detect carriers of gene deletions or duplications by quantitative gene dosage assays[10]. However, the interpretation of what constitutes a causative mutation is not always straightforward. Clearly large deletions which remove part of the coding sequence of the gene or mutations which produce a stop codon resulting in premature termination of translation of the mRNA are likely to disrupt the function of the encoded protein, but missense mutations which alter a single amino acid residue may be genuine mutations or rare variants which have no effect on protein function. Such changes must fulfil criteria such as co-segregation with the disease in the family, absence in substantial number of ethnically matched controls, and must be predicted to have a significant effect on the structure of the protein in order to be regarded as pathogenic mutations.

Linkage analysis with polymorphic markers

The first step in the identification of a disease gene has often been a linkage study, wherein the chromosomal location of the gene is established by a genome scan. Many families with multiple cases of the disease are typed with polymorphic DNA markers from each of chromosomes 1–22 (in a condition with autosomal inheritance) or from the X chromosome in X-linked disorders, until a marker is found which co-segregates with the disease. Once tightly linked markers are found, not only can the gene be identified by so-called positional cloning[1], but the markers can be used for indirect diagnosis even before the gene in question has been cloned. Once the allele of the polymorphism which is linked to the disease gene in a particular family has been identified, it can be used to track the inheritance of the mutant gene in that family and predict, for example, the affection status of a fetus. The pace of genome mapping and sequencing has reduced the mean interval from establishing a linkage to the cloning of the disease gene dramatically (it took 10 years to clone the Huntington's disease gene), thus reducing the need for this indirect form of diagnosis. However, the problems of mutation detection already mentioned have led to the continued use of linkage analysis in families in whom the causative mutation has not been identified. The choice of diagnostic method is sometimes dictated by resources, since if DNA samples from the relevant family members are available, it may be quicker (and cheaper) to type them with a few highly informative polymorphic markers than to scan their gene for the mutation. However, the problems associated with indirect diagnosis based on linkage analysis may render accurate diagnosis impossible.

These include the occurrence of new mutations, germinal mosaicism, meiotic recombination between the marker and the disease gene, and unavailability of samples from key family members.

Diagnosis of monogenic disorders

Confirmation of clinical diagnosis

The description of so many genes and their causative mutations has led to a change in the function of DNA diagnostics in recent years. Initially, mutation or linkage analysis was performed for families in whom an unequivocal clinical diagnosis had been made, in order to provide carrier testing or prenatal diagnosis. Increasingly, the function of DNA diagnostic laboratories is to test specific genes for mutations in order to confirm the clinical diagnosis or to provide the primary diagnosis. For example, a child with a chronic chest infection and a borderline sweat test might be confirmed as having cystic fibrosis by being shown to be homozygous for the common CF mutation; detection of a homozygous deletion in the gene for spinal muscular atrophy might lead to a decision to discontinue ventilation in an infant with a poor neuromuscular response; detection of a duplication of the *PMP22* gene on chromosome 17 would confirm a diagnosis of hereditary motor/sensory neuropathy type 1A. Thus disease taxonomy is increasingly being defined at the molecular level. This presages a much greater role for genetic diagnosis in the future, as the number of genes and mutations associated with human disease continues to expand.

Carrier testing and presymptomatic diagnosis

Carrier testing is now possible for a wide range of autosomal recessive disorders, but is generally only requested if there is a family history of the condition and if it is relatively common in the population in question. Examples are cystic fibrosis in Europeans and North Americans, and β-thalassaemia in certain Mediterranean populations. This requires knowledge either of the causative mutation in the family, or at least of those mutations which are common in the particular population. In the latter case, the carrier risk can be substantially reduced, but not eliminated, if the consultand does not have any of the common mutations. Carrier testing for X-linked recessive conditions such as Duchenne muscular dystrophy and haemophilia also requires knowledge of the mutation for an accurate diagnosis, although linkage analysis can be helpful if the mutation is unknown. Presymptomatic diagnosis or predictive testing for a number of adult onset disorders is also now possible for patients who are at risk because of their family history and who may wish to know whether they carry the relevant mutation. This may be in order to plan life decisions such as whether to have children and, if so, whether to undergo prenatal diagnosis, or simply to eliminate the uncertainty. The disease may be incurable, such as Huntington's disease or early onset familial Alzheimer's disease. In other cases, useful preventative action may be taken, such as surgical removal of polyps for familial adenomatous polyposis. This form of genetic testing is more complex and challenging in terms of ethical considerations and patient care, and requires clearly defined protocols and careful genetic counselling[11].

Prenatal diagnosis

Prenatal diagnosis by DNA analysis of a chorionic villus sample (CVS) taken in the 11th to 12th week of pregnancy is now routine for a large number of monogenic disorders, provided that the mutation in the affected family member is known, or that linkage analysis can determine whether the fetus has inherited the allele of the polymorphism which is linked to the mutant gene. The fact that sufficient DNA can be extracted directly from the CVS, and that the diagnosis is generally (but not always) performed by PCR, means that a result can often be provided to the family within a few days, and a first trimester termination of the pregnancy can be carried out if necessary. The molecular work-up of the family should precede the pregnancy in order to establish whether an informative prenatal test is possible. Regrettably this is not always the case, since cogitation does not always precede procreation. The CVS should also be tested to exclude even a small amount of contamination with maternal cells, since the sensitivity of PCR may lead to a false negative or false positive diagnosis as a result of amplification of a mutant or wild-type maternal allele.

Pre-implantation diagnosis

The triad of technical developments of *in vitro* fertilisation (IVF), PCR and fluorescent *in situ* hybridisation (FISH) have created the possibility of carrying out pre-implantation genetic diagnosis (PGD). The fertilised embryo is biopsied at about the 8-cell stage, with one or two cells being removed for molecular or cytogenetic analysis.[12]. Since a single cell should contain the full genetic complement of the embryo, any specific sequence can in principle be amplified by PCR and analysed for the relevant mutation. Similarly, FISH can be carried out to sex the embryo (in X-linked disorders), or to test for chromosomal abnormalities. The PGD process is time-consuming, expensive and limited by the success rates of the IVF procedure. However, there is a continuing demand for PGD from the significant minority of patients for whom termination of pregnancy is not acceptable. Molecular PGD is associated with a significant risk of diagnostic failure because of the perfect sensitivity required for PCR from a single cell, or of misdiagnosis as a result of PCR contamination from the high number of PCR cycles required to amplify a single molecule to detectable levels. Thus a PCR assay for a deletion of an exon of a particular gene, such as the dystrophin gene in Duchenne muscular dystrophy, could easily produce a false negative (*i.e.* normal) result in an affected embryo if any of the reagents were contaminated with either genomic DNA or a very small amount of the relevant PCR amplification product[13]. The use of linked highly polymorphic microsatellite markers is helpful in this context, since they provide a rather specific DNA profile of the identity of the biopsied cell. In spite of these difficulties, successful PGDs have been performed by PCR, with one group reporting the birth of 5 unaffected children after PGD for cystic fibrosis[14]. FISH currently appears to be a more reliable technique in a PGD setting, and will be used increasingly to detect chromosomal aneuploidy in cases of recurrent miscarriage of conventional pregnancies. Systematic monitoring of outcomes and error rates for PGD, and publication of these data, should be required at this rather experimental stage of the technique.

Population screening

Most DNA diagnosis is provided for individuals and families with a history of a particular genetic disorder, but there has been considerable debate and research regarding the introduction of population screening for the most common disease genes in certain populations. There are several models for this, including screening for common β-globin gene mutations, which has led to a reduction in the incidence of β-thalassaemia in Southern Europe (reviewed by Modell[15]), and antenatal screening for cystic fibrosis[16], which has been associated with a 65% reduction in the incidence of CF in the city of Edinburgh[17]. The logic of attempting to reduce the incidence of life-threatening, incurable disorders is strong, but there are numerous practical challenges involved. These include decisions as to the appropriate age and circumstances under which to screen, the cost of screening for sufficient mutations to reduce the carrier risk greatly, the cost of provision of adequate genetic counselling for those who test positive, and concerns about the insurance consequences for carriers. The technical difficulties in screening for multiple mutations will probably be resolved with the advent of new DNA technology (see below), but each screening proposal will have to be carefully assessed in terms of logistics, cost, and ethical concerns.

Genetic diagnosis of common, complex disorders

Most common disorders such as diabetes and cancer are regarded as non-genetic, in the sense that there is often only one case of the disease in a family, and even if there is more than one case, no clear pattern of Mendelian inheritance can be observed. However, epidemiological studies of risks to relatives and of disease concordance in monozygotic versus dizygotic twins have shown that genes exist which confer moderate disease susceptibility on the individual, but that the presence of several such genes, in addition to an environmental trigger, is probably required to produce the disorder. Even for common diseases, there are sometimes a subset of cases which are associated with mutations in rare single genes of high penetrance. For example, mutations in the *BRCA1* gene confer a high risk of breast and ovarian cancer, and in the Presenilin-1 gene of Alzheimer's disease. However, these mutations can generally be recognised by the early onset of the disease and the presence of a strong family history. They are, therefore, considered to be monogenic disorders, and are dealt with using diagnostic and counselling procedures already developed for disorders such as Huntington's disease. However, the majority of cases of those common diseases for which there is epidemiological evidence of a genetic component, such as diabetes and asthma, will probably turn out to be associated with mutations in multiple genes, and these genes will be relatively common in the general population. One example of this is the apolipoprotein gene, the ApoE4 allele of which is associated with an increased risk of Alzheimer's disease (reviewed by Pericak-Vance and Haines[4]). The expectation is that many more such genes will soon be identified, and that this knowledge will be applied to disease diagnosis and classification, to stratification of drug therapy, and to disease prevention, since a predetermined genetic susceptibility could lead to an altered life style which reduces the environmental component of risk[18]. Such information is likely to create an enormous demand for genetic testing in the future, which will be a new paradigm for DNA diagnostics in terms of the scale of sample throughput, the implications for those who test positive, and the method by which the information is conveyed to the patient.

The technical challenge

The major challenge which diagnostic laboratories face at present is to find the causative mutation in the relevant gene in each family referred for diagnosis. Although some disease genes have a simple mutational profile, with most cases being accounted for by one or a few mutations, for many the profile is complex. There is also the problem of genetic heterogeneity, in which mutations in more than one gene produce the same disease. In hereditary non-polyposis colorectal cancer, for example, the mutations are more or less evenly distributed between the *MLH1*

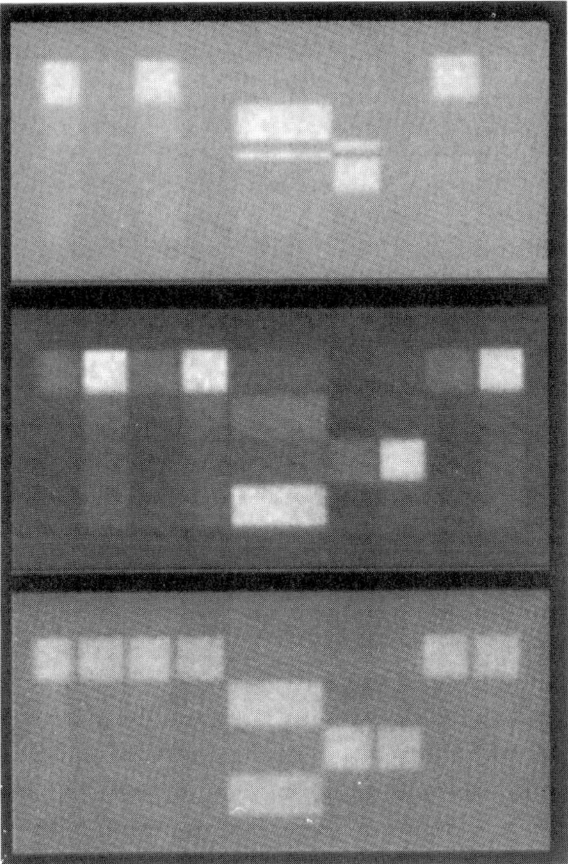

Fig. 2 Mutation screening of the CF gene using the chip. The chip in the upper panel has been hybridised with fluorescently labelled DNA from a sample homozygous for normal sequence, whereas the chip in the middle panel was hybridised with DNA homozygous for the R553X mutation. The lower panel is from a heterozygote, with both the normal and the mutant probes lighting up. (This figure was printed as part of Figure 3 in Cronin MT, Fucini RV, Kim SM, Masino RS, Wespi RM, Miyada CG. Cystic fibrosis mutation detection by hybridisation to light-generated DNA probe arrays. *Hum Mutat* 1996; 7: 244–55 and is reproduced here by kind permission of John Wiley and Sons Ltd, the copyright holder.)

and *MSH2* genes, both of which must, therefore, be screened to find the mutation in the patient. The mutation scanning methods developed thus far are labour-intensive and require considerable technical skill. There is a great need for a robust, sensitive and automated method which would allow the mutation in any gene of interest to be found in one or two days. The answer may lie in the microchip: not the electronic chip of the computer, but a chip upon which are bound tens of thousands of short oligonucleotide probes in ordered arrays of high density. Fluorescently-labelled target DNA (from the patient) is then hybridised to the chip, and will bind to complementary DNA sequences on it. Scanning of the chip with a confocal fluorescent microscope produces an image which provides information as to the nucleotide sequence of the target[19]. There are two types of configuration for this approach. The simpler of these, termed a mutation array[20], involves the synthesis of two parallel probe sets for a given disease gene, one of which is complementary to many wild-type gene sequences and the other to the corresponding sequences of many known mutations. The use of this approach to detect a mutation in the cystic fibrosis gene is shown in Figure 2. The other, more complex, configuration is a 'tiling array' in which the entire sequence of the gene, or a part of the gene, is reconstructed as sets of overlapping probes which contain all possible variations of the nucleotide sequence. In principle, the use of tiling arrays allows the entire coding sequence of a disease gene to be scanned for all possible mutations in a single experiment. The method is, therefore, an immensely sophisticated form of allele-specific hybridisation, which was developed by Wallace[21]. The equipment required for this approach comes with the modest price tag of about US$ 200 000, and chips for only a few genes are available as yet. Also, the sensitivity and accuracy of the method will have to be assessed in extensive diagnostic trials. However, the approach is exciting, is based on solid theoretical principles, and has already been applied to limited mutation scanning of the CF and *BRCA1* genes[20,22].

The other major challenge facing DNA diagnostic laboratories will be to develop high throughput methods for the detection of susceptibility genes for the common disorders. Thus far, the challenge has been diversity, with many different kinds of tests being done on small numbers of samples. For many laboratories, the highest throughput has been for the Fragile X syndrome, with up to perhaps 100 samples being tested in any one laboratory per month. If genetic tests for conditions such as diabetes and asthma seem likely to provide useful information for the diagnosis of susceptibility or choice of therapy, the demand will be on a scale without precedent in diagnostic laboratories, and will require much more automation than has been necessary hitherto. These tests are likely to involve distinguishing between two alleles which differ by a single base change, and several such tests may be necessary to define the susceptibility profile of a patient. At least one highly automated method is already available for this purpose, known as the TaqMan assay. This involves synthesis of two different oligonucleotide probes, each of which is complementary to one of the two alleles. Each probe is labelled at its 5′ end with a different fluorescent dye (the reporter), and both probes are tagged with a 'quencher' molecule which masks the signal from the reporter. During the PCR the probes bind to their complementary sequences on the products of the PCR reaction, and the reporter dye is then removed from the probe by the 5′-exonuclease activity of the *Taq* DNA polymerase enzyme[23]. This release of the reporter from the proximity of the quencher results in a progressive increase in fluorescence throughout the PCR reaction, and the relative intensity of the two dyes is read at the end of the reaction by a scanner. The associated software converts these signals to a genotype for the two alleles (AA, AB or BB). The PCR reactions are set up in 96-well plates, and once this has been done the only further action required from laboratory staff is to insert the plate into the scanner. In this way, thousands of samples can be typed in a single day. The limitation of TaqMan is that only one sequence variant or polymorphism can be typed in the assay; further technical developments will be required for the simultaneous analysis of multiple variants.

Delivery of diagnostics to the patient

The identification of the genes for most of the common monogenic disorders created an almost instant demand by clinical geneticists for the provision of DNA diagnostic services. This provision evolved rapidly from a research-based operation by the laboratories involved in mapping and cloning the genes, into a professional routine diagnostic service with an ethic based on patient care and quality control rather than on the acquisition of original data. Thus a new profession was created, with its own set of goals and challenges. Three broad principles are apparent in the delivery of this service to the patient.

Benefit to the patient or family

The primary consideration in the decision on whether to establish DNA testing, either for mutations in a particular gene or for a polymorphic allele which confers an increased risk for a disorder, must be whether the test will be of benefit to the patient or their family. This may seem obvious, but it is a principle which is easy to forget in the excitement of an intellectual recognition of the research potential of screening, for example, breast cancer patients for mutations in the *BRCA1* gene. In this instance, the detection of a mutation will confer a risk of a contralateral breast tumour, and of ovarian cancer. This should prompt questions about what practical support can be offered to such patients in terms of long-term screening for additional tumours, and possible prophylactic measures such as tamoxifen therapy or mastectomy. There will also be important consequences for the patient's offspring, who may require genetic counselling and testing for the mutation. In the context of susceptibility genes for common disorders, there seems little point in testing an asymptomatic individual for the ApoE4 allele, for example, until we have reached the point where the knowledge that he or she is at increased risk of developing Alzheimer's disease is of some use in prevention or early treatment of the condition. Thus the goal must be to provide the patient or family with information which they can use for constructive decision making, rather than to augment the publications list of the investigator.

Provision of a high quality laboratory service

This is of course a requirement for all types of diagnostic services, but the added challenges in the context of DNA diagnostics are the very rapid pace of gene discovery, which creates instant (and justifiable) demand for a new test, and of PCR technology, which allows one to set up the new test within days of the latest issue of a journal arriving in the library or on the Internet. This laudable enthusiasm must be tempered with sufficient restraint to permit reflection on the value of the test to the patient (see above), and on a technical evaluation of the accuracy of whatever assays are being put in place. There are also much broader requirements for a professional laboratory service, which should include: (i) direction of the laboratory by staff with appropriate qualifications and experience; (ii) structured internal quality control for checking of raw data and reports; (iii) strong emphasis on adequate staff training; and (iv) participation in an external quality assessment scheme. These standards can be encouraged by national laboratory accreditation schemes, such as Clinical Pathology Accreditation UK (Ltd) in the UK. In the US, the Task Force on Genetic Testing has recommended the creation of a genetics speciality for proficiency testing[24]. The development of good laboratory services is helped greatly by the establishment of strong national professional organisations, such as the Clinical Molecular Genetics Society in the UK, which organise best-practice workshops, and issue detailed guidelines on testing for specific diseases. Such organisations can also have an important role in ensuring adequate provision of a full range of services nationally, and as a source of information for the clinician on which services are available.

Linkage to genetic counselling

In the context of the monogenic disorders, where rare genes confer high risks of developing the disorder on family members and pregnancies, the provision of DNA diagnostics has generally been tightly linked to clinical genetics services which provide pre- and post-test counselling to families. This is of fundamental importance to service provision, in order to ensure that tests are only carried out if necessary and appropriate, that the often complex information on the implications of the test result are understood by the counsellor, and that this information is conveyed to the patient or family with clarity and with sympathy. A very effective model for the provision of genetic services is the integration of clinical and laboratory services within one department or centre, but if this is not possible, there should at least be a close and structured working relationship between the clinical and laboratory components of the service. Direct linkage to counselling and critical analysis of the need for a test are more difficult to achieve in countries in which many of the laboratory services are private institutions which are operated for profit[24]. In the UK, the Advisory Committee on Genetic Testing has issued a code of practice on the provision of genetic tests directly to the public[25] which recommended that this should be limited to determination of carrier status for recessive disorders, and that suppliers should facilitate genetic counselling as part of the testing procedure.

What is less clear is how we will deal with genetic information which relates to susceptibility to common disorders. This information differs from the bulk of our clinical experience with monogenic disorders in at least two respects. First, there is the question of volume, since potentially very large numbers of patients will be tested, and the labour-intensive process of genetic counselling by trained counsellors will not be sustainable. Secondly, the genotype obtained from the test will generally not be of overwhelming significance to either the patient or their family, as is the case with many of the tests for monogenic disorders, such as a positive predictive test for Huntington's disease. Rather, it will be one of several factors used by the physician to make the diagnosis and to decide on the course of treatment, and may confer quite modest risks of developing the same disease to the offspring of the patient. These differences have been used to argue that we can cut the link between genetic testing and counselling in the common disorders, and that risk factors based on DNA can be treated in the same way as other risk factors such as blood pressure or cholesterol concentration[18]. However, there are real dangers here which will have to be taken into account and provided for. First among these is ignorance of the principle that the test must provide the individual and the physician with useful information which will lead to more accurate diagnosis, disease prevention, or appropriate therapy – we already have sufficient evidence of reflex box-ticking by junior medical staff on test request forms to realise the potential difficulties here. Secondly, there will often be consequences for the offspring. If, for example, knowledge of ApoE4 status helped to define drug response in Alzheimer's disease, and the patient were found to be homozygous, the probability of homozygosity in his or her offspring would be increased and, therefore, also their risk of developing the condition at a much earlier age than the general population. Thus there will have to be a mechanism by which a consensus is reached about the situations in which the test should be applied, and a clear understanding of its implications. This information will then have to be communicated effectively to the appropriate medical speciality. A considerable degree of will power and organisation will be required in order to avoid the potentially dire consequences of haphazard and indiscriminate use of genetic information.

What began, little more than a decade ago, with an occasional prenatal diagnosis for sickle cell anaemia or thalassaemia is now a substantial profession which provides diagnostic information for hundreds of genetic disorders. It is poised for a major expansion into testing for DNA sequence variants, which will have important consequences for both diagnosis and treatment of common diseases. There are major technical challenges to be met, but perhaps more difficult will be the task of ensuring that the new information is applied appropriately for patient care.

Acknowledgements

I thank David Ellis for providing Figure 1, and am grateful to John Wiley & Sons Ltd for permission to reproduce part of Figure 3 in Cronin *et al.*[20] as Figure 2.

References

1 Weatherall DJ. *The New Genetics and Clinical Practice,* 3rd edn. Oxford: Oxford University Press, 1991

2 Cooper DN, Ball EV, Krawczak M. The human gene mutation database. *Nucleic Acids Res* 1998; **26**: 285–7

3 Cordell HJ, Todd JA. Multifactorial inheritance in type 1 diabetes. *Trends Genet* 1995; **11**: 499–504

4 Pericak-Vance MA, Haines JL. Genetic susceptibility to Alzheimer's disease. *Trends Genet* 1995; **11**: 504–8

5 Grompe M. The rapid detection of unknown mutations in nucleic acids. *Nat Genet* 1994; **5**: 111–7

6 Cotton RGH, Rodrigues NR, Campbell RD. Reactivity of cytosine and thymine in single-base-pair mismatches with hydroxylamine and osmium tetroxide and its application to the study of mutations. *Proc Natl Acad Sci USA* 1988; **85**: 4397–401

7 Rowley G, Saad S, Giannelli F, Green PM. Ultrarapid mutation detection by multiplex, solid-phase chemical cleavage. *Genomics* 1995; **30**: 574–82

8 Bates G, Lehrach H. Trinucleotide repeat expansions and human genetic disease. *Bioessays* 1994; **16**: 277–84

9 Chamberlain JS, Gibbs RA, Ranier JE, Nguyen PN, Thomas CT. Deletion screening of the Duchenne muscular dystrophy locus via multiplex DNA amplification. *Nucleic Acids Res* 1988; **16**: 11141–56

10 Yau SC, Bobrow M, Mathew CG, Abbs S. Accurate diagnosis of carriers of deletions and duplications in Duchenne/Becker muscular dystrophy by fluorescent dosage analysis. *J Med Genet* 1996;**33**:550–8

11 Craufurd D, Tyler A, on behalf of the UK Huntington's Prediction Consortium. Predictive testing for Huntington's disease: protocol of the UK Huntington's Prediction Consortium. *J Med Genet* 1992; **29**: 915–8

12 Handyside AH, Delhanty JD. Preimplantation genetic diagnosis: strategies and surprises. *Trends Genet* 1997; **13**: 270–5

13 Holding C, Bentley D, Roberts R, Bobrow M, Mathew C. Development and validation of laboratory procedures for preimplantation diagnosis of Duchenne muscular dystrophy. *J Med Genet* 1993; **30**: 903–9

14 Asangla AO, Ray P, Harper J *et al.* Clinical experience with preimplantation genetic diagnosis of cystic fibrosis (ΔF508). *Prenat Diagn* 1996; **16**: 1370–42

15 Modell B. Cystic fibrosis screening and community genetics. *J Med Genet* 1990; **27**: 475–9

16 Brock DJH. Heterozygote screening for cystic fibrosis. *Eur J Hum Genet* 1995; **3**: 2–13

17 Cunningham S, Marshall T. Influence of five years of antenatal screening on the paediatric cystic fibrosis population in one region. *Arch Dis Child* 1998; **78**: 345–8

18 Bell J. The new genetics in clinical practice. *BMJ* 1998; **316**: 618–20

19 Lipshutz RJ, Morris D, Chee M *et al.* Using oligonucleotide probe arrays to access genetic diversity. *BioTechniques* 1995; **19**: 442–7

20 Cronin MT, Fucini RV, Kim SM, Masino RS, Wespi RM, Miyada CG. Cystic fibrosis mutation detection by hybridisation to light-generated DNA probe arrays. *Hum Mutat* 1996; **7**: 244–55

21 Conner BJ, Reyes AA, Morin C, Itakura K, Teplitz RL, Wallace BR. Detection of sickle cell β-globin allele by hybridization with synthetic oligonucleotides. *Proc Natl Acad Sci USA* 1983; **80**: 278–82

22 Hacia JG, Brody LC, Chee MS, Fodor SPA, Collins FS. Detection of heterozygous mutations in BRCA1 using high density oligonucleotide arrays and two-colour fluorescence analysis. *Nat Genet* 1996; **14**:441–7

23 Livak KJ, Marmaro J, Todd JA. Towards fully automated genome-wide polymorphism screening. *Nat Genet* 1995; **9**: 341–2

24. Holtzman NA, Shapiro D. Genetic testing and public policy. *BMJ* 1998; **316**: 852-6

25 Advisory Committee on Genetic Testing. *Code of Practice and Guidance on Human Genetic Testing Services Supplied Direct to the Public.* London: Health Departments of the United Kingdom, 1997

Disease taxonomy – monogenic muscular dystrophy

Jacques S Beckmann

URA CNRS 1922 – Généthon, Evry, France

The field of the autosomal recessive progressive muscular dystrophies has clarified significantly following the recent elucidation of the genetic and molecular etiology of a number of these entities. These studies illustrate how genetics provides a rationale and objective basis for a new, refined nosology. Furthermore, whereas most of these studies point towards the pivotal role played by a number of structural proteins – all directly or indirectly associated with dystrophin – a calpain protease was shown to be involved in the Réunion-type limb girdle muscular dystrophy. This discovery raises the issue of whether these mechanisms are all part of one and the same pathway or of distinct pathophysiological pathways (structuropathy versus enzymopathy) leading to similar phenotypes. Finally, all of these diseases are considered as classical monogenic traits. Some findings suggest, however, that epistatic interactions have been overlooked and that the inheritance models could be slightly more complex. These results are discussed in light of the coming challenges of the identification of genes underlying complex multifactorial traits.

Whereas once controversial, there is no longer much debate over the notion that virtually all diseases have a genetic component, the identification of which holds many promises for public health. It is thus not surprising that much of the current efforts in human genetics are directed towards this goal. One should not forget, however, that the availability of increasingly powerful analytical methods in this research area is a recent event. And yet, even if it amounts to only the tip of the iceberg, considerable success has already been achieved. It is therefore legitimate, from time to time, to look back and attempt to draw conclusions, which may be of relevance to the coming challenges.

The genetic bases of inherited diseases span from simple monogenic entities (such as Duchenne's muscular dystrophy, cystic fibrosis, thallassaemia, Tay-Sachs disease) to complex multifactorial traits (*e.g.* diabetes, obesity, hypertension, schizophrenia or Alzheimer's disease). It is generally agreed that the former lend themselves to 'simple' genetic analyses. This is essentially due to the fact that they are thought to be

Correspondence to: Dr Jacques S Beckmann, URA CNRS 1922 – Généthon, 1 Rue de l'Internationale, 91000 Evry, France

'Mendelising', *i.e.* that the different genotypic classes fall into discrete, easily recognisable phenotypic categories. It is, thus, no surprise that the application of the powerful arsenal of genetic methodologies to the study of these diseases has already led to the identification of the molecular aetiology of several hundred disease loci. This identification has been greatly facilitated by the fact that analyses of the co-segregation in these families of linked markers and the disease phenotype allows one to define a candidate interval delineated by true recombinant boundaries. This situation is to be contrasted with the latter group of traits, the complex traits, which are controlled both by genetic and non-genetic factors, and for which there is a broad, continuous distribution of phenotypes.

Recent observations suggest, however, that reality might be more subtle, *i.e.* that the demarcation between these two groups may be more diffuse. We shall review here some of the progress made towards the elucidation of the genetic aetiology of autosomal progressive muscular dystrophies, particularly of limb girdle muscular dystrophy (LGMD), emphasizing the impact on nosology, or what will be referred here as reverse medicine, as well as the 'take home messages' that may be pertinent to the study of complex traits.

The autosomal progressive muscular dystrophies

The term muscular dystrophy (MD) refers to a group of myogenic debilitating disorders, the most notorious and most frequent one being the recessive X-linked Duchenne/Becker MD. The latter are caused by a dystrophin-deficit. Analysis of the inheritance patterns of other MD types allows the distinction of two groups of progressive MD, the autosomal dominant (AD) and recessive (AR) forms. We shall base our discussion on the work done in the study of the autosomal recessive entities.

These AR progressive muscular dystrophies constitute a genetically and clinically heterogeneous group of diseases of low prevalence (10^{-5})[1], in which there is a progressive wasting of skeletal muscle fibres, characterised by a necrosis-regeneration dystrophic pattern. The concept of LGMD was introduced in their classical paper by Walton and Nattrass[2], describing cases whose cardinal features were 'onset usually late in the first decade, or in the second or third decade but sometimes in middle age, commencement of muscular weakness in either the shoulder or pelvic girdle, transmission usually via an autosomal recessive gene and a relatively slow course which nevertheless leads to severe disablement or often death before the normal age'.

Despite the fact that the first description of such patients was made by Erb back in 1884[3], the legitimate nosological existence of an LGMD diagnosis remained, until recently, challenged or severely criticized by some leading authors (*e.g.* Bradley[4] and Brooke[5]). This is essentially due to the fact that there were no consensual distinctive diagnostic criteria that enable one to specifically recognize and diagnose these patients. The latter are characterised by a number of parameters, none of which is specific to these disorders. As a result, the LGMD diagnosis is often firstly, a diagnosis of exclusion and, secondly, plagued with misascertainment[6,7].

Actually, during the last 15 years, the field of AR progressive muscular dystrophies was developing rapidly. Examination of muscle biopsies from patients demonstrated that dystrophin was normally present. Because of our inability, prior to the advent of genetic diagnosis, to separate objectively these clinical entities one from another and thus to individualise specific pathognomic traits, and, in view of the large overall phenotypic overlap, these diseases were arbitrarily lumped together under the common denomination of LGMD2s[8].

It is thus no surprise that even within the community working on these diseases this situation has led to heated debates on the adequacy of the proposed nomenclature, some arguing that it only adds to the confusion and that the term LGMD should be reserved exclusively to the phenotype initially described by Erb[3]. Fortunately, the identification of the underlying molecular mechanisms contributes to clarify this semantic and nosologic problem, as recently dempnstrated in an ENMC workshop on the lgmds[9]. Genetics provides thus a rationale and objective basis for a new, refined nosology ('reverse medicine').

From disease to gene

Gene identification, even for 'simple' monogenic traits, is still not a simple task. There are recipes and common principles. Yet, each story has its own idiosyncrasies and surprises. No wonder then that different strategies were followed to identify the causative genes of AR progressive muscular dystrophies. These different approaches will be briefly summarised here.

The discovery of a group of patients in the southern part of the Réunion Island, whose age of onset, pattern of muscular involvement, and rate of evolution[9] fitted well with the original description of the juvenile form given by Erb[3] prompted the genetic study that led to the primary localisation of the LGMD2A locus to the long arm of chromosome 15[10], subsequently confirmed in other families[11,12]. This genetic mapping provided the demonstration of the legitimacy for the existence of this clinical entity, thereby settling this dispute. Subsequently, a long and tedious positional cloning (reviewed by Beckmann *et al*[13]) effort led eventually to the identification of mutations within the CAPN3 gene encoding the muscle-specific calcium-activated neutral protease, calpain 3[14]. Note that despite the fact that its cDNA had been known since 1989[15], it had never been considered *a priori* as a functional candidate gene. Furthermore, the identification of LGMD2A patients carrying two null type mutation alleles demonstrated that lack of calpain 3 activity can be pathogenic.

During this period, and since the discovery of the role of dystrophin in 1987[16] in Duchenne and Becker MD, exceedingly elegant biochemical studies led to the characterisation of the oligomeric complex of dystrophin-associated proteins (DAP)[17–19], a number of which were also found to be missing in DMD patients[20,21]. The implication of these proteins in the pathophysiology of progressive muscular dystrophies was soon suspected upon the demonstration of a deficiency of one of these proteins, a 50 kDa glycoprotein, subsequently termed α-sarcoglycan, in MD patients[22]. The corresponding gene thus became an attractive candidate gene for this disease. No wonder that this promptly led to the demonstration of the role of the α-sarcoglycan gene in some, yet, and this should be emphasized, not in all α-sarcoglycanopathies[23,24]. For instance, the gene segregating in a set of families of Tunisian origin is distinct from the α-sarcoglycan gene, as it maps onto a different chromosome[25,26], even though patients from these families show an α-sarcoglycan deficit. In other words, mutations elsewhere on the genome can also lead to such a deficiency, hence the need to distinguish between primary and secondary α-sarcoglycanopathies[27].

The identification of the role of the α-sarcoglycan gene was done following a candidate gene strategy: specific antibodies for α-sarcoglycan had been produced in the course of the biochemical studies of the DAP complex[28]; partial peptide sequences were determined; the corresponding cDNA clones were isolated from a rabbit cDNA library, and eventually mapped onto human chromosome 17q21. The demonstration of the co-segregation in one multiplex family of markers specific for this chromosomal region (including an intragenic marker) with the disease phenotype, was followed by the uncovering of the causative mutations in the α-sarcoglycan gene[23].

Yet, this left unresolved the issue of α-sarcoglycanopathies in families, which could not be accounted for by mutations in the chromosome 17q21 region. When antibodies against the other sarcoglycans became available, immunocytochemical diagnoses on biopsies of patients with α-sarcoglycan-deficiencies revealed that these proteins were also missing or strongly reduced in these patients. This observation led to the hypothesis that a primary deficiency in anyone of the components of the sarcoglycan complex could result in a similar phenotype, and the sarcoglycan genes became prime candidates genes for these diseases[29,30].

The β-sarcoglycan gene, mapping to chromosome 4q12, was incriminated in another MD entity following a combined conventional positional cloning and candidate gene strategy[31,32]. Concurrent to the demonstration of the role of the β-sarcoglycan gene, the γ-sarcoglycan gene mapping to chromosome 13q12 was demonstrated to segregate with the disease in the 'Tunisian' dystrophies[33]. Finally, when a 'new' 35 kDa sarcoglycan was identified (δ-sarcoglycan[34]), screening of affected families for which the role of the other sarcoglycans had been excluded[35], led finally to the detection of mutation(s) within this gene mapping to chromosome 5q33[36].

All the considered biochemical and genetic observations led to the proposal that these proteins, through their association with dystrophin and the dystroglycan complex, provide a continuous link from the cytoskeletal F-actin to the extracellular matrix, and stabilise the sarcolemma and lipid bilayer during muscle contraction. Muscular dystrophies could, therefore, be considered at first sight as diseases of the dystrophin-glycoprotein complex[37–39].

Occasionally, genetic studies can also give insights into the biochemical structures of the encoded proteins. Consider the sarcoglycanopathies, caused by a deficiency in one of a set of proteins known to be part of a higher order structure, the DAP complex. The fact that all sarcoglycanopathies documented to date have always been found to follow an AR inheritance, indicated the stoichiometry of these proteins in this complex, *i.e.* it allowed one to rule out the existence of homopolymers, as the latter would have been expected to occasionally lead to AD traits[40].

Thus, altogether five distinct MD genes were identified in a relatively short time span: one through a conventional positional cloning strategy, two through a functional candidate strategy, and the remaining two, by a combination of a candidate gene and a positional cloning strategies. These five genes, encoding respectively for one of the four sarcoglycans and calpain 3, still can not account for all AR progressive MDs. There remain families for which neither of these disease loci can explain the phenotype. Interestingly, immunocytochemical staining on muscle biopsies of these patients for the sarcoglycans or dystrophin is normal. Thus we are not facing sarcoglycanopathies. And we know that at least three other genes remain to be identified. These were mapped, respectively, to chromosomes 2p (LGMD2B[41,42]), 17q (LGMD2F[43]) and 9q (LGMD2H[44]). The LGMD2B locus on 2p has also an interesting twist. As a matter of fact, it was the second LGMD2 locus to be mapped. Interestingly, the distal Miyoshi type of MD was also mapped to the same region[45], raising the question of whether these two entities might be allelic or only syntenic. This issue was recently settled by genetic analyses, as two recent reports described within two large highly consanguineous kindreds, a set of clinically different siblings that were found to be geno-identical for the LGMD2B/Miyoshi interval on chromosome 2p[46,47]. These observations strongly suggest that the

Miyoshi and 'LGMD2B' myopathies are not only allelic, but that the same pathological mutation could, depending on additional unknown factors, lead to one or another clinical condition. Following the intensive concerted positional cloning efforts[49,50], the chromosome 2p locus eventually revealed its secrets. The identification of dysferlin mutations settled the responsibility of the DYSF locus in the etiology of both diseases. Furthermore, these molecular studies also unambiguously demonstrated that the same mutation could lead to the first or second condition[49,51], suggesting the possible contribution of additional factors in the determination of the final phenotype. Like dystrophin and the sarcoglycans, dysferlin is also localized in the myofibers' periphery[52], although its relationship – if any – to the DAP complex is still unclear.

Our discussion on AR progressive MDs does not terminate here. During the quest for AD lgmds, Minetti *et al.*[53] reported patients from 2 families with mutations in caveolin 3, a protein localised in sarcolemmal caveolae. Thus LGMD1C became the first AD lgmd locus to be recognized. Subsequently, McNally *et al.*[54] reported a MD patient that was a homozygous carrier for a missense substitution. The trait was thus presumably recessively inherited. Apparently and not unexpectedly, one could thus according to this study, encounter either an AD or AR lgmd depending on the nature and impact of the mutation.

Yet another disease may also yield a lgmd-like phenotype. Haravuori *et al.*[55] suggested, upon the mapping in an extended Finnish pedigree of the gene responsible for the AD tibial MD to 2q31, that the underlying locus was the titin gene. Some patients in this pedigree exhibit, however, a lgmd-like proximal phenotype. Genetic analyses suggested that these patients carry the presumed titin mutation in double dose. If this were the case, presence of the mutation in single or double dose would lead respectively to a distal AD or proximal AR MD.

Finally, the validation of the causal nature of the recognised sequence variants should be discussed. This is indeed a relevant concern for rare monogenic disorders. But it may, for other reasons, also apply to complex traits. In other words, what tools do we have to discriminate a (possibly rare private) neutral polymorphism from a pathogenic variant. This is particularly pertinent for sequence changes that lead to missense or even synonymous substitutions. It is not our purpose here to discuss this at length, but just to bring two illustration to document this difficulty.

The necessity to demonstrate the pathogenic nature of a variant works both ways. While searching for pathogenic calpain 3 mutations, a number of neutral variants were encountered, some of which were even frequently seen on control chromosomes[56]. This example emphasizes the necessity to carefully assess the pathogenic nature of the encountered variants, as one could otherwise be misled to erroneous inference or diagnosis. But even what may seem innocuous needs not be so. In our quest for calpain 3 mutations. In our quest for calpain 3 mutations, one consanguineous family from La Réunion was encountered that showed, as expected, a homozygous LGMD2A haplotype. Yet sequencing the entire coding sequence of calpain 3 only revealed one supposedly 'silent' mutation that affected the third base of a glycine codon. It was only through the demonstration that this variant altered the normal splicing pattern of the calpain 3 mRNA and was effectively equivalent to a null-type mutation that its pathogenic identity was confirmed[48]. This illustration still represents an easy case, as one is on safe ground if one can demonstrate, at the molecular level, the invalidating nature of a mutation. Settling this issue in the case of rare missense variants may be much more difficult.

Reverse medicine: from gene to disease

The knowledge of the genetic etiology opens the door to a new type of question pertaining to molecular physiology. It is now possible to challenge and relate the activities of specific MD genes to particular types of physical exercise or movements. This type of approach is likely to yield valuable information insofar as it will provide clues as to when and where these genes' activities are solicited. But it will also be of immediate relevance to patients' care management, by identifying those movements that are either least or most deleterious to their particular state.

As stated above, setting a progressive muscular dystrophy diagnosis was not a straightforward matter. As a matter of fact, it is through the elucidation of their molecular etiology that these different entities have been and continue to be, individualised. The availability of molecular, and genetic means allows one to objectively recognise and individualise the various AR myopathies. It is now possible to establish detailed phenotype/genotype relationships, to specify the patterns of involvement of different muscle groups, to assess possible correlations between the nature or site of the mutation and the resulting phenotype, to compare their respective clinical features, to recognise the phenotypic nuances, and to precise the nosological boundaries of each one of these similar, yet different entities[9,49–51]. The capacity to rely on an unbiased diagnosis, should allow the establishment of specific discriminating features, and eventually of a precise definition of their nosology. The first results from such 'retrospective' clinical analyses suggest that the most significant diagnostic criterion is the specific pattern of muscle involvement[49–52], though the extent of overlap of the clinical manifestations of these different clinical entities is still unknown. It is thus no longer justified to treat 'LGMD' cases in bulk. Slowly, but surely, the nosological boundaries of these entities start to clarify.

To sum up, owing to genetics, we know nowadays already of at least eight independent loci that can contribute to an AR progressive muscular dystrophy phenotype. Besides the important new leads in clinical analyses, this knowledge provides new accurate diagnostic tools such as a panel of antibodies directed against the proteins involved in these disorders (or not inconceivably, eventually even specific enzymatic assays, such as for calpain 3). This has already had a very consequential impact on nosology. The availability of these immunodiagnostic and molecular genetic tools allows one to distinguish two groups of AR progressive MDs: those that lead to loss of the sarcoglycan protein(s) and are caused by mutation(s) in one of the corresponding genes – the sarcoglycanopathies (for an excellent review see Ozawa *et al.*[39]), and those where the sarcoglycan complex is preserved. This refinement opens the way to an understanding of the pathophysiological bases of each of these respective disorders.

The pathophysiology of muscular dystrophy: enzymopathy or structuropathy?

In parallel to the developments recounted above, the uncovering of the genetic etiology of the AD–LGMDs has also progressed. It suffices to say here that five distinct loci have been incriminated so far[9], only one of which was, as recalled earlier, recently identified. It encodes the structural protein caveolin 3, which colocalises with dystrophin at the muscle fibre membrane, and is thus apparently also part of the DAP complex[53]. Hence, this AD–LGMD is to be included among the diseases of the dystrophin glycoprotein complex[37–39]. Is the situation for LGMD2A the same? It should be remembered that almost all other MD genes identified thus far encode for a structural protein, each inactivating one or another component of the (intra- or extracellular) cytoskeletal infrastructure (see Beckmann & Fardeau[52] for a more detailed list), and so constitute the group of structuropathies.

Does this apply to calpain 3? The validation of the involvement of calpain 3 in LGMD2A is the first demonstration of an enzymatic rather than a structural protein defect causing a progressive muscular dystrophy. Apparently, the sarcoglycan and calpain 3 genes seem to account for two distinct biochemical processes, yet mutations in any of these genes lead to a similar syndrome. And in all these cases, loss of function can be pathogenic. The precise function of the calpain protease as well as the nature of its biological substrate(s) remains unknown.

The incrimination of a protease in a myopathy raises intriguing questions. Why and how would a protease deficiency result in an overall similar necrosis/regeneration dystrophic pattern and clinical phenotype?

For all we know, the structural proteins constituting the DAP complex are not affected in calpain 3-deficient patients; there is no evidence in favour of the involvement of this protease in the maturation process of any of these proteins (F. Leturcq, unpublished observations). Could it be, that besides its proteolytic activity, calpain 3 also assumes a structural role? Sorimachi *et al.*[54] showed that it was able to bind titin, a structural protein extending over half a sarcomere unit length. Whereas this is compatible with a structural role, it certainly does not prove it. Thus it is still unclear if and how calpainopathies fit in the group of the dystrophin-glycoprotein complex diseases[37–39].

The recent suggestion that titin mutations may lead to a MD, which in heterozygous or homozygous carriers leads respectively to a distal or proximal selectivity, supports this contention. We still remain with the open question of whether we are facing two distinctive pathophysiological mechanisms that can cause similar phenotypes, or whether there is a functional or hierarchical link between these proteins and this calpain? In other words, is calpain 3, in some manner as yet unbeknown to us, related to the same physiological pathway as the other structural proteins? And, if it is, does it act in parallel or in series to these other gene products? If the latter is true, does it act upstream or downstream of their action? Providing an answer to these questions is crucial for the understanding of the pathophysiology of these muscular dystrophies and for the potential development of new therapies. Further analyses will be required to clarify these points.

Determination of expression territories

Numerous strategies can be used to elucidate the biochemical and physiological function of a defined protein. We would like to emphasize the utility of the determination of spacio-temporal expression patterns of the corresponding genes during human (and mouse) embryonic development to identify functional cues[55]. Although this approach has been applied to both the sarcoglycans and calpain 3, we shall review here only the latter study.

Calpain 3 is the first, and still the only mammalian, member of the family of calcium-dependent cysteine-proteases for which a connection was established with a defined phenotype. Despite the fact that calpains have been invoked as major players in a number of different biological processes, little is actually known of their real functions. The same holds for calpain 3. It is also still unclear how a protease deficiency could lead to a MD. Calpain 3 expression was reported to be specific to the skeletal muscle[15]. Examination of the spacio-temporal developmental expression pattern revealed that this gene is transcribed throughout human embryonic development in a variety of different tissues, including the heart and smooth muscle[56]. It should be emphasized, however, that the presence of calpain 3 RNA in the heart was unexpected considering the absence of any recorded clinical cardiac sign in LGMD2A patients[49,50]. As a matter of fact, the latter is a criterion for the exclusion of the LGMD2A diagnosis. One has, therefore, to reconcile the fact that this gene is active in tissues, at least as judged by the presence of the corresponding mRNA, and presumably fulfilling a function, yet patients having no calpain 3 activity are considered to be free of heart symptoms. What these and other similar results demonstrate, is that the specificity of gene expression need not to be restricted to the 'clinical target' tissue, even for loss of function mutations.

The following hypotheses can be forwarded to explain this apparent conflict. The gene may be transcribed but no functional protein is made (the availability of antibodies should soon enable to answer this point). Alternatively, this gene is either not playing an essential role, or the absence of calpain 3 is compensated for, *e.g.* by an activity from a redundant pathway.

In addition, transcription of this gene is also subject to tissue-specific alternative splicing[56,65], or transcription initiation (Hérasse *et al.* [65]; Ma *et al.*[57]). One still needs to demonstrate whether the various calpain 3 RNA isoforms all lead to the synthesis of the corresponding translation products. However, if confirmed, the presence of differentially expressed protein isoforms further complicates the analyses and unraveling of this gene's function(s); these elements will need to be integrated before we can come to grip with calpain 3's biological role(s) and the pathophysiological consequences of a deficiency in this gene's product(s).

Qualitative or quantitative phenotype

It is generally assumed that for monogenic diseases the alternative genotypes fall into distinct qualitative phenotypic classes. A closer examination of the AR progressive muscular dystrophies shows that this is too simplistic a representation. Patients manifest a wide phenotypic diversity, as measured, for instance, by age at onset, rate of evolution or severity. Consider the sarcoglycanopathies. Some patients belonging to this group show a very severe early onset, Duchenne-like MD[58,59] while others have inadvertently come as adults to the attention of the clinicians because of apparently non-pathogenic 'muscle fatigue'[60]. What these and other studies tell us, is that often minor or conditionally silent mutations can easily be overlooked, as compared to the more easily recognizable pathogenic mutations. For primary α-sarcoglycan deficiencies, the factor determining the severity of the affliction has been shown to be the specific nature of the mutation(s) carried by the patients[59].

The same holds for calpain-deficient (LGMD2A) patients, where a wide distribution of age at onset and rate of disease progression – depending both at first sight on the type of CAPN3 mutation – has also been reported[49,50,56]. In addition, age at onset appears to be a poor prognostic factor of the degree of severity of the disease (*e.g.* Dinçer *et al.*[61]).

Furthermore, although there is in general less intra- than interfamilial variability, there are cases where geno-identical siblings differ greatly with respect to these criteria, albeit the patterns of affected muscles are the same[58,60,62–64]. To complicate the matter further, we still need to account for the fact that the same mutation, even within the same pedigree, can lead to two supposedly distinct MD phenotypes[46,47].

In this situation, one is clearly confronted with monogenic traits showing a continuous phenotypic distribution with respect to many of these parameters. The reasons, therefore, are still unknown but could reflect the fact that the phenotypic effect of a particular LGMD2 allele can be modified by the nature of the second mutated allele, by genetic factors in the vicinity or at other loci, or even by non-genetic factors (for a further discussion on these possibilities, see below).

The power of isolated populations and the Réunion paradox

It is common knowledge that the investigation of genetically heterogeneous traits can greatly benefit from the study of isolated populations. Thus, in the course of our genetic study of lgmd, three defined (highly) consanguineous sets of families were examined, the Old Order Amish families[65,66], Basque families from Spain[67], and pedigrees from a small community on the Réunion Island[10]. They were all considered *a priori* to represent homogenous genetic isolates reflecting each a unique founder mutation. Yet, an unsuspected genetic or allelic heterogeneity was demonstrated in each case, respectively[14,67–69]. Though this may not be representative of the general situation, these results nevertheless clearly illustrate the caution that needs to be taken in the interpretation of genetic studies of presumed isolated populations.

Remember the inherent diagnostic difficulties and potential genetic heterogeneity of AR progressive muscular dystrophies. It is thus evident that in this context, the initial incentive to launch a genetic study of lgmd stemmed from the discovery of a cluster of lgmd families on the Réunion island: based on genealogical studies, these pedigrees were considered to form a megafamily with multiple consanguineous links, that originated from a single common ancestor who was among the first settlers of this island in the 17th century. Hence, this megafamily was presumed to represent a clinically and genetically homogenous set. That the study of these families eventually led to the primary localisation of the LGMD2A locus (the first lgmd gene to be mapped) lent further credence to the hypothesis that these patients belong to a relatively young and genetically homogenous isolate[10].

The discovery in this population of at least six distinct segregating LGMD2A haplotypes came as a surprise[68]. The subsequent demonstration of an equivalent number of different calpain 3 mutations[14,48] established the allelic heterogeneity. A similar observation was reported in the Basque calpain-deficient cohort[67]. These observations create an enigma, which we termed the 'Réunion paradox', namely, how can one explain the presence of this many mutations for such rare conditions in presumably genetically homogenous inbred isolates. This phenomenon is apparently definitely not restricted to calpainopathies. There are a number of observations reporting a high frequencies of mutations in defined genes in isolated inbred populations from small geographic areas (e.g. Bach et al.[70], Rodius et al.[71], Heinisch et al.[72], Zlotogora et al.[73]).

Complexifying monogenic inheritance

Several alternative explanations[14,69,73] can be considered to account for the 'Réunion paradox'. Yet eventually the most plausible and parsimonious of these invokes an extension of the digenic inheritance initially demonstrated for *retinitis pigmentosa* by Kajiwara *et al.*[74]. In this model, mutations in two unlinked genes are required to result in a phenotype. The founder effect in the Réunion island population could thus be on this second, as yet unidentified locus.

We would like to stress that this concept is likely to represent common situations in genetics[14,75], explaining among others incomplete penetrance (even for recessive traits) and variable expressivity, as well as the failure to reproduce human genetic diseases in animal models.

An attractive alternative explanation that could also account for these phenomena was recently brought to our attention. Prevalence estimates, based on genetic epidemiological studies, only yield average estimates, the variance around which could be very significant. In other words, what might *a priori* appear as unexpected, based on the general prevalence value, may be viewed completely differently if one considers the variance, *i.e.* if one allows, in selected populations, for outliners. Yet, even then it seems hard to reconcile the fact that this phenomenon has been observed repeatedly in the Arab Israeli population for a number of different AR traits (see Zlotogora *et al.*[73] for more references), and that in the Basque LGMD2A population, which also represents an isolated genetic niche with high degree of consanguinity, at least 5 different calpain 3 mutations have been identified[67].

Predictions of the digenic model

Though not demonstrated as yet for calpainopathies, the digenic model is likely to have many ramifications that extend beyond this particular case. It may be worth briefly restating a few of the predictions of the digenic model (for a more detailed discussion see Beckmann[76]). First of all, the population frequency of mutant alleles should be higher than the frequency estimated based on disease prevalence only. Second, several independent mutations are expected (remember, because of the conditional pathogenicity of the mutations, this gene is not subject to the usual counter-selection pressures). Our observations for the LGMD2A population, namely the identification of close to 100 distinct pathogenic calpain 3 mutations, and the fact that three out of four chromosomes carry an as yet unidentified mutation, support this inference. Finally, as the two 'disease loci' are unlinked, we expect some families in the general population to have asymptomatic carriers, *i.e.* individuals with pathogenic mutation(s) yet no clinical symptoms. Owing to the low overall prevalence of calpainopathies, this may be hard to prove with current methodologies, for CAPN3. Yet, this is exactly what has been observed in a number of other diseases[77–82], where siblings that were geno-identical at the disease locus could, nevertheless, differ not only with respect to the expressivity, but even penetrance of the disease.

Other support for the digenic model comes from mouse knock-out studies. There is, indeed, a growing body of evidence where a second, or sometimes third, gene needs to be inactivated to result in a phenotype reproducing a human disease (for review, see Wynshaw-Boris[83] and Cooke *et al.*[84]). Furthermore, the elegant mouse study of Rozmahel *et al.*[85] on cystic fibrosis transmembrane conductance regulator deficiency – another 'monogenic' trait – can also be cited as additional support of a departure from classical monogenic inheritance, wherein the phenotype of a well-known monogenic trait is modulated by epistatic interactions determined by the genetic context in which it occurs.

Extension of the digenic model

An important, as yet undocumented, prediction of the flexibility endowed by the digenic model is that the partners involved in this partial functional overlap may vary from one tissue to another. In this case, a single mutation may result in distinct diseases depending on the identity of the susceptibility locus. Phenotypes could thus vary as a function of tissue-specificity of the predisposing alleles. Thus, the 'one mutation – several diseases' concept could well be explained by a digenic inheritance model.

A pertinent example of such a scenario is the Miyoshi-LGMD2B case reported above (for further examples, see Beckmann[76]). Although this is still a speculative proposal, further dissection of the genetic bases of numerous inherited disorders will undoubtedly clarify this point. Extrapolating this to polygenic traits, the same allele could be involved in the etiology of diverse diseases, affecting distinct tissues in different individuals.

Conclusions

The elucidation of the etiology underlying the AR progressive muscular dystrophies, has already impacted in both the diagnosis and in laying the foundations for an understanding of the pathophysiology of these diseases. It now becomes possible to provide better patient care management and to offer, hopefully in the not too distant future, significant therapeutic perspectives. Meanwhile, the nosological boundaries of these entities can be clarified (reverse medicine).

As mentioned earlier, every gene identification exercise has its own specificities and, from each of one them, there are messages to be learnt. A different story might have led to different, equally pertinent, lessons. The road towards deciphering the content of our genome is long, non-repetitive and full of the unexpected: quite a fascinating adventure for the curious genome tourist. Yet one of the main messages one ought to draw, is that the simplistic linear view of 'one gene – one function – one phenotype' is far from representative of the whole story, even for a single gene. A better understanding of normal and pathophysiology will require both the uncovering and integration of possible epistatic interactions, and an improved capacity to discern subtle phenotypic nuances. In most instances, unless we position a gene's product in its complex multidimensional biological context, and identify all partners with which it directly or indirectly interacts, we will not be able to get a clear picture of its function. In other words the challenge is now to place each gene product into its physiological pathway(s). The latter may definitively often involve complex structured gene and functional networks.

Finally, the simple dichotomy between purely monogenic and multifactorial traits has proven to be a very powerful working model in human genetics. But, as we learn more about inherited characters, we realise that this model may need to be revisited. Besides the classical 'monogenic' traits, there may be a continuum of traits that are not so monogenic after all or only conditionally monogenic. Furthermore, even within specific genes, there may be mutations that behave as typical monogenic characters, while the expression of other allelic variants may require the interaction with specific genetic – or even environmental – contexts.

Hopefully, on the other side of this continuum, we may find out that a number of so-called complex multifactorial traits, once unravelled, may turn out to represent a genetically heterogeneous mixture of less complex genetic entities. Not everything needs to be as simple or as complex as they appear.

Acknowledgements

This work was supported by the Association Française contre les Myopathies (AFM). I thank all my colleagues and students over many years for sharing with me this exhilarating experience. I also thank Drs J.L. Guenet, J.A. Machado and J.V. Neel for stimulating discussions and suggestions.

References

1 Emery AEH. Population frequencies of inherited neuromuscular diseases – a world survey. *Neuromusc Disord* 1991; **1**: 19–29

2 Walton JN, Nattrass FJ. On the classification, natural history and treatment of the myopathies. *Brain* 1954; **77**: 169–231

3 Erb W. Ueber die 'Juvenile Form' der progressiven Muskelatrophie ihre Beziehungen zur sogehannten Pseudohypertrophie der Muskeln. *Dtsch Archiv Klin Med* 1884; **34**: 467–519

4 Bradley WG. The limb-girdle syndromes. In: Vinken PJ, Bruyn GW, Ringel SP, eds. *Diseases of Muscle. Handbook of Clinical Neurology*. Amsterdam: North-Holland, 1979: 433–69

5 Brooke MH, ed. A clinician's view of neuromuscular diseases. Baltimore:Williams & Wilkins, 1977

6 Bushby KMD. Diagnostic criteria for the limb-girdle muscular dystrophies: report of the ENMC consortium on limb-girdle muscular dystrophies. *Neuromusc Disord* 1995; **5**: 71–4

7 Beckmann JS, Bushby K. Advances in the molecular genetics of autosomal recessive progressive muscular dystrophies. *Curr Opin Neurol* 1996; **9**: 389–93

8 Bushby KMD, Beckmann JS. Report of the 30th and 31st ENMC International Workshops on the limb-girdle muscular dystrophies - proposal for a new nomenclature. *Neuromusc Disord* 1995; **5**: 337–43.

9 Beckmann J, Brown R, Muntoni F et al. Workshop report: the 66th/67th ENMC sponsored workshop ñ the limb-girdle muscular dystrophies. *Neuromuscular Disorders* 1999; In press.

10 Beckmann JS, Richard I, Hillaire D *et al.* A gene for limb-girdle muscular dystrophy maps to chromosome 15 by linkage. *C R Acad Sci III* 1991; **312**: 141–8

11 Passos-Bueno M-R, Richard I, Vainzof M *et al.* Evidence of genetic heterogeneity for the adult form of limb-girdle muscular dystrophy following linkage analysis with 15q probes in Brazilian families. *J Med Genet* 1993; **30**: 385–7

12 Young K, Foroud T, Williams P *et al.* Confirmation of linkage of limb-girdle muscular dystrophy, type 2, to chromosome 15. *Genomics* 1992; **13**: 1370–1

13 Beckmann JS, Richard I, Broux O *et al.* Identification of muscle-specific calpain and β-sarcoglycan genes in progressive autosomal recessive muscular dystrophies. *Neuromusc Disord* 1996; **6**: 455–62

14 Richard I, Broux O, Allamand V *et al.* A novel mechanism leading to muscular dystrophy: mutations in calpain 3 cause limb girdle muscular dystrophy type 2A. *Cell* 1995; **81**: 27–40

15 Sorimachi H, Imajoh-Ohmi S, Emori Y *et al.* Molecular cloning of a novel mammalian calcium-dependant protease distinct from both m- and mu- type. Specific expression of the mRNA in skeletal muscle. *J Biol Chem* 1989; **264**: 20106–11

16 Monaco AP, Kunkel LM. Cloning of the Duchenne/Becker muscular dystrophy locus. *Adv Hum Genet* 1988; **17**: 61–98

17 Campbell KP, Kahl SD. Association of dystrophin and an integral membrane glycoprotein. *Nature* 1989; **338**: 259–62

18 Ervasti JM, Campbell KP. Membrane organization of the dystrophin-glycoprotein complex. *Cell* 1991; **66**: 1121–31

19 Yoshida M, Ozawa E. Glycoprotein complex anchoring dystrophin to sarcolemma. *J Biochem (Tokyo)* 1990; **108**: 748–52

20 Ervasti JM, Ohlendieck K, Kahl SD, Gaver MG, Campbell KP. Deficiency of a glycoprotein component of the dystrophin complex in dystrophic muscle. *Nature* 1990; **345**: 315–-9

21 Matsumura K, Nonaka I, Tomé FMS *et al.* Mild deficiency of dystrophin-associated proteins in Becker muscular dystrophy patients having in-frame deletions in the rod domain of dystrophin. *Am J Hum Genet* 1993; **53**: 409–16

22 Matsumura K, Tomé FMS, Collin H *et al.* Deficiency of the 50K dystrophin-associated glycoprotein in severe childhood autosomal recessive muscular dystrophy. *Nature* 1992; **359**: 320–2

23 Roberds SL, Leturcq F, Allamand V *et al.* Missense mutations in the adhalin gene linked to autosomal recessive muscular dystrophy. *Cell* 1994; **78**: 625–33

24 Romero NB, Tomé FMS, Leturcq F *et al.* Genetic heterogeneity of severe childhood autosomal recessive muscular dystrophy with adhalin (50 kDa dystrophin-associated glycoprotein) deficiency. *C R Acad. Sci. III* 1994; **317**: 70–6

25 Ben Othmane K, Ben Hamida M, Pericak-Vance MA *et al.* Linkage of Tunisian autosomal recessive Duchenne-like muscular dystrophy to the pericentromeric region of chromosome 13q. *Nat Genet* 1992; **2**: 315–7

26 Ben Othmane K, Speer MC, Stauffer J *et al.* Evidence for linkage disequilibrium in chromosome 13-linked Duchenne-like muscular dystrophy (LGMD2C). *Am J Hum Genet* 1995; **57**: 732–4

27 Jeanpierre M, Carrié A, Piccolo F *et al.* From adhalinopathies to alpha-sarcoglycanopathies. An overview. *Neuromusc Disord* 1996; **6**: 463–5

28 Roberds SL, Anderson RD, Ibraghimov-Beskrovnaya O, Campbell KP. Primary structure and muscle-specific expression of the 50-kDa dystrophin-associated glycoprotein (adhalin). *J Biol Chem* 1993; **268**: 23739–42

29 Mizuno Y, Noguchi S, Yamamoto H *et al.* Selective defect of complex in severe childhood autosomal recessive muscular dystrophy muscle. *Biochem Biophys Res Commun* 1994; **203**: 979–83

30 Ozawa E, Yoshida M, Suzuki A, Mizuno Y, Hagiwara Y, Noguchi S. Dystrophin-associated proteins in muscular dystrophy. *Hum Molec Genet* 1994; **4**: 1711–6

31 Lim LE, Duclos F, Broux O *et al.* β-Sarcoglycan: characterization and role in limb-girdle muscular dystrophy linked to 4q12. *Nat Genet* 1995, **11**: 257–65

32 Bönnemann CG, Modi R, Noguchi S *et al.* Mutations in the dystrophin-associated glycoprotein β-sarcoglycan (A3b) cause autosomal muscular dystrophy with disintegration of the sarcoglycan complex. *Nat Genet* 1995; **11**: 266–73

33 Noguchi S, McNally EM, Ben Othmane K *et al.* Mutations in the dystrophin-associated glycoprotein γ-sarcoglycan in chromosome 13 muscular dystrophy. *Science* 1995; **270**: 819–22

34 Nigro V, Moreira ES, Piluso G *et al.* Autosomal recessive limb-girdle muscular dystrophy, LGMD2F, is caused by a mutation in the δ-sarcoglycan gene. *Nat Genet* 1996; **14**: 195–8

35 Passos-Bueno M-R, Moreira ES, Vainzof M, Marie SK, Zatz M. Linkage analysis in autosomal recessive limb-girdle muscular dystrophy (AR LGMD) maps a sixth form to 5q33-34 (LGMD2F) and indicates that there is at least one more subtype of AR LGMD. *Hum Molec Genet* 1996; **6**: 815–20

36 Nigro V, Piluso G, Belsito A *et al.* Identification of a novel sarcoglycan gene at 5q33 encoding a sarcolemmal 35 kDa glycoprotein. *Hum Mol Genet* 1996; **5**: 1179–86

37 Campbell KP. Three muscular dystrophies: loss of cytoskeleton extracellular matrix linkage. *Cell* 1995; **80**: 675–9

38 Worton R. Muscular dystrophies: diseases of the dystrophin-glycoprotein complex. *Science* 1995; **270**: 755–6

39 Ozawa E, Noguchi S, Mizuno Y, Hagiwara Y, Yoshida M. From dystrophinopathy to sarcoglycanopathy: Evolution of a concept of muscular dystrophy. *Muscle Nerve* 1998; **21**: 421–38

40 Beckmann JS. Genetic studies and molecular structures: the dystrophin associated complex. *Hum Mol Genet* 1996; **5**: 865–7

41 Bashir R, Strachan T, Keers S *et al.* A gene for autosomal recessive limb-girdle muscular dystrophy maps to chromosome 2p. *Hum Mol Genet* 1994; **3**: 455–7

42 Bashir R, Keers S, Strachan T *et al.* Genetic and physical mapping at the limb-girdle muscular dystrophy locus (LGMD2B) on chromosome 2p. *Genomics* 1996; **33**: 46–52

43 Moreira ES, Vainzof M, Marie SK, Sertié AL, Zatz M, Passos-Bueno MR. New LGMD locus (LGMD2G) mapped to 17q11-q12. *Am J Hum Genet* 1997; **61**: 151–6

44 Weiler T, Greenberg CR, Zelinski T *et al.* A gene for autosomal recessive limb-girdle muscular dystrophy in Manitoba Hutterites maps to chromosome region 9q31-q33: evidence for another LGMD locus. *Am J Hum Genet* 1998; **63**: 140–7

45 Bejaoui K, Hirabayashi K, Hentati F *et al.* Linkage of Miyoshi myopathy (distal autosomal recessive muscular dystrophy) locus to chromosome 2p12-14. *Neurology* 1995; **45**: 768–72

46 Weiler T, Greenberg CR, Nylen E *et al.* Limb-girdle muscular dystrophy and Miyoshi myopathy in an aboriginal Canadian kindred map to LGMD2B and segregate with the same haplotype. *Am J Hum Genet* 1996; **59**: 872–8

47 Illarioshkin S, Ivanova-Smolenskaya IA, Tanaka H *et al.* Clinical and molecular analysis of a large family with three distinct phenotypes of progressive muscular dystrophy. *Brain* 1996; **119:** 1895–909

48 Richard I, Beckmann JS. How neutral are synonymous codon mutations? *Nat Genet* 1995; **10**: 259

49 Liu J, Aoki M, Illa I, Wu C *et al.* Dysferlin, a novel skeletal muscle gene, is mutated in Miyoshi myopathy and limb girdle muscular dystrophy. *Nature Genet* 1998; **20**: 31–6

50 Bashir R, Britton S, Strachan T *et al.* A gene related to Caenorhabditis elegans spermatogenesis factor fer-1 is mutated in limb-girdle muscular dystrophy type 2B. *Nature Genet* 1998; **20**: 37–42

51 Weiler T, Bashir R, Anderson L *et al.* Identical mutation in patients with limb girdle muscular dystrophy type 2B or Miyoshi myopathy suggests a role for modifier gene(s). *Hum Molec Genet* 1999; **8**: 871–7

52 Anderson L, Davison K, Moss J *et al.* Dysferlin is a plasma membrane protein and is expressed early in human development. *Hum Molec Genet* 1999; **8**: 855–61

53 Minetti C, Sotgia F, Bruno C *et al.* Mutations in the caveolin-3 gene cause autosomal dominant limb girdle muscular dystrophy. *Nat Genet* 1998; **18**: 365–8

54 McNally E, de Sa Moreira E, Duggan D *et al.* Caveolin-3 in muscular dystrophy. *Hum. Molec Genet* 1998; **7**: 871–7

55 Haravuori H, Makela-Bengs P, Udd B *et al.* Assignment of the tibial muscular dystrophy locus to chromosome 2q31. *Am J Hum Genet* 1998; **62**: 620–6

56 Richard I, Roudaut C, Saenz A et al. Calpainopathy-A survey of mutations and polymorphisms. *Am J Hum Genet* 1999; **64**(6):1524-40.

57 Ma H, Fukiage C, Azuma M, Shearer TR. Cloning and expression of mRNA for calpain Lp82 from rat lens: splice variant of p94. *Invest Ophthalmol Vis Sci* 1998; **39**: 454–61

58 Ben Hamida M, Fardeau M, Attia N. Severe childhood muscular dystrophy affecting both sexes and frequent in Tunisia. *Muscle Nerve* 1983; **6**: 469–80

59 Piccolo F, Roberds SL, Jeanpierre M *et al.* Primary adhalinopathy: a common cause of autosomal recessive muscular dystrophy of variable severity. *Nat Genet* 1995; **10**: 243–5

60 Carrié A, Piccolo F, Leturcq F *et al.* Mutational diversity and hot spots in the α-sarcoglycan gene in autosomal recessive muscular dystrophy (LGMD2D). *J Med Genet* 1997; **34:** 470–5

61 Dinçer P, Leturcq F, Richard I *et al.* A biochemical, genetic and clinical survey of autosomal recessive limb girdle muscular dystrophies in Turkey. *Ann Neurol* 1997; **42**: 222–9

62 McNally E, Passos-Bueno R, Bönnemann CG *et al.* Mild and severe muscular dystrophy caused by a single γ-sarcoglycan mutation. *Am J Hum Genet* 1996; **59**: 1040–7

63 Pénisson-Besnier I, Richard I, Dubas F., Beckmann JS, Fardeau M. Pseudo-metabolic expression and phenotypic variability of calpain deficiency in two siblings. *Muscle Nerve* 1998; 21(8): 1078–80

64 Richard I, Brenguier L, Dinçer P *et al.* Multiple independent molecular etiology for limb girdle muscular dystrophy type 2A patients from various geographical origins. *Am J Hum Genet* 1997; **60**: 1128–38

65 Herasse M, Ono Y, Fougerousse F *et al.* Expression and functional characteristics of Calpain 3 isoforms generated through tissue-specific transcriptional and post-transcriptional events. *Mol Cell Biol* 1999; **19**(6): 4047–55

66 Jackson CE, Strehler DA. Limb-girdle muscular dystrophy: clinical manifestations and detection of preclinical disease. *Pediatrics* 1968; **41**: 495–502

67 Urtasun M, Saenz A, Roudaut C *et al.* Limb-girdle muscular dystrophy in Guipuzcoa (Basque country, Spain). *Brain* 1998; **121**(Pt 9): 1735–47

68 Allamand V, Broux O, Richard I *et al.* Preferential localization of the limb-girdle muscular dystrophy type 2A gene in the proximal part of a 1 cM 15q15.1-q15.3 interval. *Am J Hum Genet* 1995; **56**: 1417–30

69 Allamand V, Broux O, Bourg N *et al.* Genetic heterogeneity of autosomal recessive limb-girdle muscular dystrophy in a genetic isolate (Amish) and evidence for a new locus. *Hum Mol Genet* 1995; **4**: 459–63

70 Bach G, Moskowitz SM, Tieu PT *et al.* Molecular analysis of Hurler syndrome in Druze and Muslim Arab patients in Israel: multiple allelic mutations of the IDUA gene in a small geographic area. *Am J Hum Genet* 1994; **53**: 330–8

71 Rodius F, Duclos F, Wrogemann K *et al.* Recombinations in individuals homozygous by descent localize the Friedrich Ataxia locus in a cloned 450 kb interval. *Am J Hum Genet* 1994; **54**: 1050–9

72 Heinisch U, Zlotogora J, Kafert S, Gieselmann V. Multiple mutations are responsible for the high frequency of metachromatic leukodystrophy in a small geographic area. *Am J Hum Genet* 1995; **56**: 51–7

73 Zlotogora J, Gieselmann V, Bach G. Multiple mutations in a specific gene in a small geographic area: a common phenomenon? *Am J Hum Genet* 1996; **58**: 241–3

74 Kajiwara K, Berson EL, Dryja TP. Digenic retinitis pigmentosa due to mutations at the unlinked peripherin/RDS and ROM1 loci. *Science* 1994; **264**: 1604–8

75 Van Ommen G-J. A foundation for limb girdle muscular dystrophy. *Nat Med* 1995; **1**:412–4

76 Beckmann JS. The Réunion paradox and the digenic model. *Am J Hum Genet* 1996; **59**: 1400–2

77 Cobben JM, Van der Steege, Grootscholten P, de Visser M, Scheffer H, Buys CHCM. Deletions of the survival motor neuron gene in unaffected siblings of patients with spinal muscular atrophy. *Am J Hum Genet* 1995; **57**: 805–8

78 Hahnen E, Forkert R, Marke C *et al.* Molecular analysis of candidate genes on chromosome 5q13 in autosomal recessive spinal muscular atrophy: evidence of homozygous deletions of the SMN gene in unaffected individuals. *Hum Molec Genet* 1995; **4**: 1927–33

79 Hollan S, Fujii H, Hirono A *et al.* Hereditary triosephosphate isomerase (TPI) deficiency: two severely affected brothers one with one without neurological symptoms. *Hum Genet* 1993; **92**: 486–90

80 Risch N, de Leon D, Ozelius L *et al.* Genetic analysis of idiopathic torsion dystonia in Ashkenazi Jews and their recent descent from a small founder population. *Nat Genet* 1995; **9**: 152–9

81 Mulligan LM, Eng C., Attié T *et al.* Diverse phenotypes associated with exon 10 mutations of the Ret proto-oncogene. *Hum Molec Genet* 1994; **3**: 2163–7

82 Wang CH, Xu J, Carter TA *et al.* Characterization of survival motor neuron (SMN^T) gene deletions in asymptomatic carriers of spinal muscular atrophy. *Hum Mol Genet* 1996, **5**: 359–65

83 Wynshaw-Boris A. Model mice and human disease. *Nat Genet* 1996; **13**: 259–60

84 Cooke J, Nowak MA, Boerlijst, Maynard-Smith J. Evolutionary origins and maintenance of redundant gene expression during metazoan development. *Trends Genet* 1997; **13**: 360–4

85 Rozmahel R, Wilschanski M, Matin A *et al.* Modulation of disease severity in cystic fibrosis transmembrane conductance regulator deficient mice by a secondary genetic factor. *Nat Genet* 1996; **12**: 280–7

Disease taxonomy – polygenic

William OCM Cookson

Nuffield Department of Clinical Medicine, University of Oxford, John Radcliffe Hospital, Oxford, UK

The practice of medicine depends on the recognition and classification of disease. Correct diagnosis is the cornerstone of correct treatment. The past century has seen the classification of disease move from a reliance on symptoms and signs to the use of more and more sophisticated measurements of human structure and function. However, although most diseases have now have names and schemes of classification, these names still may hide a fundamental lack of understanding of the causes of the disease.

The extraordinary progress in molecular genetics in the last 20 years now means that a complete understanding of the constitutional predisposition to disease is possible. All disease results from the interaction between adverse environmental events and constitutional (genetic) resistance or susceptibility. Genetic resistance is modified by ageing. The study of genetics is the process of linking polymorphism in the genetic material to polymorphism or variation in the function or appearance of an organism.

The extent to which this becomes clinically useful will be determined by the strength of the genetic effects influencing the disease. Oligogenic disorders, in which just a few genes are impacting on the disease, are more likely to be classifiable by genetic polymorphism than true polygenic disorders, in which a multiplicity of small effects give incremental risks of developing disease. Nevertheless, an improved understanding of the aetiology of disease will in all probability identify previously unrecognised yet distinct subsets of disease.

Correspondence to: Prof WOCM Cookson, Professor of Human Genetics, Nuffield Department of Clinical Medicine, University of Oxford, John Radcliffe Hospital, Oxford OX3 9DU, UK

The process of finding the genes that cause common disorders is only beginning, and the underlying structural DNA changes are, by and large, not yet known. It is possible to anticipate the formidable task ahead by looking to the single gene disorders.

Inborn defects in haemoglobin are the best understood of all inherited disorders, and have been studied the longest at the molecular level. Sickle cell disease is due to a single nucleotide substitution in the β-globin gene, resulting in HbS. Even though it arises from a single specific mutation, sickle cell disease has pleiotropic manifestations, with a wide range of symptoms which differ from one patient to the next.

British Medical Bulletin 1999;**55** (No. 2): 358–365

Most schoolchildren know that the sickle mutation in the heterozygous state confers protection against falciparum malaria, and that, for this reason, the mutation is common in areas of endemic malaria infection. The sickle mutation appears to have arisen on 5 different occasions in 5 different haplotypes (four African, one Indian). The severity of the disease relates to the particular haplotype which carries the HbS mutation.

Mutations in and around the β-globin gene produce β-thalassaemia, due to reduced production of β-globin. Thalassaemia trait is also protective against malaria, and is endemic in southern Europe. Over a hundred different variants in the β-globin gene are known, affecting controlling and splice sequences as well as causing changes in coding sequences.

The theme of multiple events affecting the same gene in similar ways is very likely to occur in common complex disorders. These mutations or polymorphisms will also be prevalent in the population because they have in the past conferred some protection against some adverse environmental force. They will be modified by the genetic background of the individual, as well as by changes between the modern Western environment and the past, in which infection and poor nutrition had a huge impact on individual survival.

Diabetes

Diabetes is the most common metabolic disease. Less than 100 years ago, diabetes was considered to be a single disease. It was known that patients suffered from an excess of sugar in the urine. After the identification of insulin, it was recognised that diabetes was sometimes characterised by insulin excess rather than insulin lack. Insulin excess was seen typically in adult onset disease, or non-insulin dependent diabetes (NIDDM) which is also known as type II diabetes. Juvenile onset disease was characterised by insulin lack (IDDM) and is known as type I diabetes.

Type I diabetes shows familial aggregation. Many studies have now been carried out which have attempted to localise and identify the genes which predispose to the disorder[1]. The strongest effect by far is that of the HLA class II genes, including HLA-DR and HLA-DQ. The HLA-DQ locus may be the most important. About 95% of Caucasians with type I diabetes will carry either HLA-DR3 or HLA-DR4 or both alleles. These alleles are in linkage disequilibrium with DQβ1 *0302 and DQβ1*0201 alleles, which confer a higher risk of disease that DR3 or DR4, and are thought to contribute directly to disease susceptibility. The combination of DQβ1*0302 and DQβ1*0201 carries a particularly high risk. DR2 and DQβ1*0602, which are also in disequilibrium, are protective against the disease.

These clear-cut associations have not yet explained how the HLA locus modifies the risk of disease. HLA class II restriction of the immune response to the principal diabetes auto-antigen has not been demonstrated. The alleles seem to be permissive to some external event, which has not yet been characterized.

Many other chromosomal regions are being investigated for the possible presence of IDDM genes. Not unexpectedly, given the difficulty in cloning genes of small effect, the presence of genes in some of these regions is contentious[2]. Of the non-HLA genes, the best studied is the Insulin locus on the short arm of chromosome 11. A variable number of tandem repeat polymorphism (VNTR) within this locus may actually predispose to disease[3].

How much use is this information in the classification of juvenile diabetes? HLA-DR typing is certainly of interest in unaffected siblings of the diabetic proband, and will become more relevant if strategies for the prevention of disease are developed. The insulin VNTR does not yet have a place in the classification of disease. The remaining type I diabetes loci remain unhelpful in the clinical context.

Type II diabetics are in general obese, and suffer from insulin resistance rather than insulin lack. Type II diabetes is very common. It too is strongly heritable, implying a genetic predisposition. The very high prevalence of the disease in some populations, such as Pima Indians and Australian Aborigines is thought to be a consequence of a selective advantage of type II diabetes genes in conditions of poor nutrition.

The molecular basis of the disease is not known. Studies of its molecular genetics are particularly difficult because of the late onset of the illness, and have not yet defined chromosomal regions or candidate genes that influence disease susceptibility.

A rare variant of type II diabetes has been recognised in which disease comes on in early middle life. Patients are resistant to insulin, and tend not to develop keto-acidosis. This disorder has been labelled as maturity onset disease of the young, or MODY. Apart from the early onset of illness, patients differ from type II diabetics in that they are thin. The disease is strongly familial, segregating as an autosomal dominant disorder.

Genetic studies have shown different MODY families to be linked to different chromosomal regions[4]. MODY2 was linked to chromosome 7, and has subsequently been shown to be due to mutations in the glucokinase gene. MODY1 on chromosome 20 and MODY3 on chromosome 12q are due to mutations of the transcription factors of the hepatic nuclear family (HNF-1 for MODY3 and HNF-4 for MODY1).

Genetics in these cases has identified a previously completely unexpected mechanism for the disease.

Mutations in glucokinase/MODY2 result in mild chronic hyperglycaemia, whereas MODY1 and MODY3 are characterised by severe insulin secretory defects and major hyperglycaemia associated with microvascular complications[4]. Mutations in these genes, therefore, define subtypes of diabetes with distinctive clinical courses and responses to treatment[4,5]. MODY genes do not seem to have a major role in common late-onset NIDDM.

Asthma

Asthma affects one child in seven in the UK and one adult in twelve. It is characterised by airway inflammation and intermittent airflow obstruction. Although asthma is often considered as a single disorder, it is almost certainly not one disease but many.

The most common form of asthma is allergic asthma, also known as atopic asthma: 95% of childhood asthma is atopic. The prevalence of childhood asthma rises to a peak in the early teens, and is more common and earlier in onset in boys than in girls. Even within the childhood asthma there are likely to be several unrecognised syndromes. Asthma may be mild and easily treated, or it may be chronic and unremitting, or brittle with sudden life-threatening episodes against a background of mild disease.

In adults, most cases of asthma are also atopic, but many individuals with adult-onset asthma have no evidence of underlying atopy. In these cases, cigarette smoking is often, but not invariably, a contributing cause. Other asthma syndromes, with known precipitants, are also recognised. Aspirin sensitive asthma affects 10% of adult onset asthmatics and, although aspirin ingestion leads to asthma in these patients, it does not seem to be through classical allergic pathways. Industrial asthma is well recognised. Baker's asthma, and laboratory animal worker's asthma are seen in atopic individuals, but isocyanate (a paint additive) and other industrial exposures lead to asthma through non-atopic mechanisms.

Almost all genetic studies of asthma have concentrated on classical allergic asthma. Segregation analysis has indicated the presence of major genes underlying atopy and asthma, with the expectation that these genes may be identified by positional cloning. As with type I diabetes, a number of chromosomal regions have been identified as containing genes which influence asthma and atopy[6].

Two regions have been studied in detail. The first is the human MHC. The HLA-DR genes modify the ability to respond to particular allergens[7], such as house dust mite or grass or tree pollens. The relative risk of specific allergy associated with a particular HLA-DR type is usually less than 2, and is strongly influenced by environmental events. It seems unlikely that HLA-DR typing will ever be of prognostic relevance for allergic asthma.

Stronger HLA class II associations have been seen with aspirin sensitive asthma[7] and also with industrial isocyanate-induced asthma[7] and, if these results are generally confirmed, it is possible that HLA typing may be of use in identifying individuals at risk so that exposure may be avoided.

Polymorphisms within the tumour necrosis factor genes (within the MHC) also predispose to asthma, possibly by enhancing the intensity of airway inflammation[8]. A clinical role for these polymorphisms is yet to be explored, but TNF genotypes might differentiate between atopics who get asthma and those who do not, or indicate that anti-inflammatory treatment may be required sooner rather than later.

The second region which has been well studied is on chromosome 11q13. This region contains the important candidate gene FcεRI-β (the β chain of the high affinity receptor to IgE). Linkage to this region was originally seen in families with severe atopy, and polymorphism within the gene has been associated with asthma, bronchial hyper-responsiveness and infantile eczema[9,10]. It is quite possible that genotyping at this locus will eventually define a syndrome of early onset severe allergic disease, and may anticipate a difficult clinical course.

Although the identification of other asthma genes is incomplete, the early identification of children at genetic risk of asthma seems possible. Whereas the precision of the estimate of risk is unlikely to be as accurate as for single gene disorders, the prevention of illness by environmental or other intervention in susceptible children is both feasible and desirable.

Psychiatric disorders

Certain psychiatric disorders show familial aggregation, most notably manic depressive psychosis (bipolar disorder) and schizophrenia. The search for genes underlying these syndromes has been active for over a decade, but, as yet, there are no agreed linkages or candidate gene effects. It may be argued that this suggests a polygenic background[11], with no major locus likely to influence the disease strongly enough to be of diagnostic use. Nevertheless, subsets of disease may well become apparent, just as they have with NIDDM and MODY. In the further future, the capricious response to many psychotropic drugs may also become predictable with genotypic information.

Alzheimer's disease

One person in ten above the age of 70 years in the UK has chronic memory loss, and half of these patients are thought to suffer from Alzheimer's disease (AD). A family history of dementia is strongly predictive of AD, indicating an underlying genetic predisposition.

Genetic studies have led to the identification of three genes which, when mutated, cause familial forms of AD. These genes are the β-amyloid precursor protein gene (APP), and the presenilin 1 (PS-1) and presenilin 2 (PS-2) genes[12]. Mutations in APP on chromosome 21 are a very rare cause of autosomal dominant AD. Mutations in the highly homologous genes PS-1 (on chromosome 14) and PS-2 (on chromosome 1) are both associated with an autosomal dominant early-onset disease[13,14].

In cases of late onset AD, with or without a family history, allele 4 of the apolipoprotein E (ApoE) gene increases risk for disease in a dose-dependent manner[15]. The polymorphism has been described as a susceptibility allele, because not all patients with ApoE4 will get the disease. A more accurate description would be that it is a disease gene with incomplete penetrance. 40–50% of the risk for late-onset disease has been attributed to the locus. PS-1 also seems important in late-onset AD, and may account for about half as much of the risk for AD as ApoE4[16].

APP, PS-1 and PS-2 may cause disease by influencing amyloid handling within the CNS. A plausible mechanism for ApoE4 predisposing to AD has not been established. The excess of cardiovascular deaths in AD patients with ApoE4 alleles[17] suggests that susceptibility to disease may be mediated through vascular mechanisms.

The clinical use of genetic polymorphisms in AD is not yet established. Typing for PS-1 or PS-2 polymorphism may be of diagnostic importance in an individual with early onset dementia, particularly if there is a suggestive family history. Mutations in the APP gene are very rare, and screening for them is unlikely to become part of general clinical practice.

Typing for ApoE polymorphisms may support the diagnosis of AD in a dementia of any age, but will not give a definitive diagnosis.

Although it has been suggested that the ApoE4 genotype implies a poor response to acetylcholinesterases[18], in general genotyping for any of these known polymorphisms carries no therapeutic advantage for the patient, nor yet any information that may help prevent disease in relatives.

Inflammatory bowel disease

Inflammatory bowel disease (IBD), which includes the diatheses of ulcerative colitis and Crohn's disease, also shows strong familial aggregation. Recently, four chromosomal regions (on chromosomes 3, 7, 12 and 16) have been identified which show linkage to IBD[19–21]. The locus in chromosome 16 seems specific to Crohn's disease[20,21]. The relative ease with which these linkages have been replicated suggests that genes of strong effect underlie the syndrome, with the corollary that the loci will contain alleles of diagnostic importance.

Conclusions

In this early stage of the search for genes underlying common disorders, it is uncertain which truly polygenetic diseases will be improved in their classification by genetics. Nevertheless, diabetes and Alzheimer's disease already appear quite differently to clinicians than they did only 10 years ago, and it is likely that the clinical approach to IBD and asthma will soon also be changed as a result of genetic information. Even if most cases of a disease are due to interactions between multiple polygenes of small effect, it seems inevitable that new distinct clinical entities will emerge from every disease with a substantial familial component.

References

1 Todd JA, Farrall M. Panning for gold: genome-wide scanning for linkage in type 1 diabetes. *Hum Mol Genet* 1996; 5: 1443–8
2 Lernark Å, Ott J. Sometimes it's hot, sometimes it's not. *Nat Genet* 1998; **19**: 213–4
3 Bennett ST, Lucassen AM, Gough SC *et al.* Susceptibility to human type 1 diabetes at IDDM2 is determined by tandem repeat variation at the insulin gene minisatellite locus. *Nat Genet* 1995; **9**: 284–92
4 Velho G, Froguel P. Maturity-onset diabetes of the young (MODY), MODY genes and non-insulin-dependent diabetes mellitus. *Diabetes Metab* 1997; **23 Suppl 2**: 34–7
5 Frayling TM, Bulamn MP, Ellard S *et al.* Mutations in the hepatocyte nuclear factor-1 alpha gene are a common cause of maturity-onset diabetes of the young in the UK. *Diabetes* 1997; **46**: 720–5
6 Daniels SE & Bhattacharyya S, James A *et al.* A genome-wide search for quantitative trait loci underlying asthma. *Nature* 1996; **383**: 247–50
7 Moffatt MF, Cookson WOCM. The genetics of specific allergy. *Monogr Allergy* 1996; **33**: 71–97
8 Moffatt MF, Cookson WOCM. Tumour necrosis factor haplotypes and asthma. *Hum Mol Genet* 1997; **6**: 551–4
9 Hill MR, Cookson WOCM. A new variant of the β subunit of the high-affinity receptor for immunoglobulin E (FcεRI-β E237G): associations with measures of atopy and bronchial hyper-responsiveness. *Hum Mol Genet* 1996; **5**: 959–62
10 Cox HE, Moffatt MF, Faux JA et al. Association of atopic dermatitis to the beta subunit of the high affinity immunoglobulin E receptor. *Br J Dermatol* 1998; **138**: 182–7
11 Crow TJ. Current status of linkage for schizophrenia: polygenes of vanishingly small effect or multiple false positives? *Am J Med Genet* 1997; **74**: 99–103

12 Goate AM. Molecular genetics of Alzheimer's disease. *Geriatrics* 1997; **52**: S9–12
13 Alzheimer's Disease Collaborative Group. The structure of the presenilin I (S182) gene and identification of six novel mutations in early onset AD families. *Nat Genet* 1995; **11**: 219–22
14 Cruts M, Hendriks L, Van Broeckhoven C. The presenilin genes: a new gene family involved in Alzheimer disease pathology. *Hum Mol Genet* 1996; **5**: 1449–55
15 Higgins GA, Large CH, Rupniak HT, Barnes JC. Apolipoprotein E and Alzheimer's disease: a review of recent studies. *Pharmacol Biochem Behav* 1997; **56**: 675–85
16 Wragg M, Hutton M, Talbot C. Genetic association between intronic polymorphism in presenilin-1 gene and late-onset Alzheimer's disease. Alzheimer's Disease Collaborative Group. *Lancet* 1996; **347**: 509–12
17 Olichney JM, Sabbagh MN, Hofstetter CR *et al.* The impact of apolipoprotein E4 on cause of death in Alzheimer's disease. *Neurology* 1997; **49**: 76–81
18 Poirier J, Delisle MC, Quirion R *et al.* Apolipoprotein E4 allele as a predictor of cholinergic deficits and treatment outcome in Alzheimer disease. *Proc Natl Acad Sci USA* 1995; **92**:12260–4
19 Satsangi J, Parkes M, Louis E *et al.* Two stage genome-wide search in inflammatory bowel disease provides evidence for susceptibility loci on chromosomes 3, 7 and 12. *Nat Genet* 1996; **14**: 199–202
20 Hugot JP, Laurent-Puig P, Gower-Rousseau C *et al.* Mapping of a susceptibility locus for Crohn's disease on chromosome 16. *Nature* 1996; **379**: 821–3
21 Ohmen JD, Yang HY, Yamamoto KK *et al.* Susceptibility locus for inflammatory bowel disease on chromosome 16 has a role in Crohn's disease, but not in ulcerative colitis. *Hum Mol Genet* 1996; **5**: 1679–83

Pharmacogenetics

C Roland Wolf* and Gillian Smith†

**Imperial Cancer Research Fund, Molecular Pharmacology Unit and †Biomedical Research Centre, Ninewells Hospital and Medical School, Dundee, UK*

Inter-individual variability in drug response is a major clinical problem. Adverse drug reactions (ADRs) are common, are responsible for a number of debilitating side effects following drug therapy and are a significant cause of death. It is now clear that much of the observed variability in drug response has a genetic basis, arising as a result of genetically-determined differences in drug absorption, disposition, metabolism or excretion. The best characterised pharmacogenetic polymorphisms are those within the phase I cytochrome P450 family of drug metabolising enzymes. One of these enzymes, CYP2D6 (debrisoquine hydroxlyase), metabolises one-quarter of all prescribed drugs and is inactive in 6% of the Caucasian population. Individuals at risk of developing ADRs as a result of genetically-determined variation in genes such as CYP2D6 can now be identified using DNA-based tests. A detailed knowledge of the genetic basis of individual drug response is potentially of major clinical and economic importance and could provide the basis for a rational approach to drug prescription. This would have significant benefits for human health.

Build me newer molecules,
O my Soul –
As the swift seasons roll
Let each new compound
Safer than the last
Avoid the reactions observed in the past
Till all at length are free
From vexing idiosyncrasy

The Pharmacologic Principles of Medical Practice (1954)

Correspondence to: Prof C Roland Wolf, Imperial Cancer Research Fund, Molecular Pharmacology Unit, Ninewells Hospital and Medical School, Dundee DD1 9SY, UK

This quotation, cited by Krantz and Carr in *The Pharmacologic Principles of Medical Practice* in 1954[1], illustrates how rudimentary our knowledge of variability in drug response was some 40 years ago. These authors did, however, appreciate the significant clinical problem posed by adverse drug reactions.

In contrast to the above rather pessimistic view, most modern drugs as they are commonly used are safe. However, there remains enormous individuality in drug response; side effects are common which, although usually not life threatening, can be unpleasant and debilitating to the patient. Every year in the UK as many as 20,000 serious adverse drug reactions are reported; fatal ADRs have recently been identified as the fourth leading cause of death in the US, after heart disease, cancer and stroke[2].

Individuality in response to therapeutic drugs is, therefore, of major clinical and economic importance. Such variability can be manifest either as a lack of therapeutic benefit when drug clearance is too rapid or by a spectrum of adverse drug reactions which, in certain extreme cases, can be lethal, for example when impaired drug clearance leads to accumulation and toxicity. The ability to predict or avoid such adverse events would allow drugs to be prescribed in a manner which is of maximum benefit to the patient and which would obviate the need to use alternative drugs or to adjust the optimum dose empirically based on patient response. This has obvious implications for the prescribing practises of general practitioners, would avoid unnecessary drug costs and reduce avoidable hospitalisation.

At present, this is an idealistic view. However, our increased understanding of the factors which determine therapeutic response and the recent major technological advances in analytical genetic analysis are moving towards making the screening of patients for sensitivity to certain types of drug therapy a reality. Such applications are currently being routinely applied in the pharmaceutical industry in the design and analysis of clinical trials. Individual variability in drug response is also an increasingly important aspect of the drug regulatory and registration processes.

Whether an individual patient will respond or will not respond to a particular drug treatment and whether they will experience any or severe side effects can be determined by a wide range of different mechanisms. Some of these are directly related to metabolism or arise as a consequence of drug/drug interactions when several drugs are prescribed concomitantly. Adverse drug reactions due to drug/drug interactions are clearly avoidable. However, there is increasing evidence that a significant proportion of these mechanisms may be genetically determined – the ability to routinely use genetically-based methods to predict individual response to drug treatment is now becoming a realistic goal. The genetic basis for individuality in response to drugs has been trivially termed 'pharmacogenetics'. This is a 'trivial term' because it trivialises a fundamental aspect of the evolutionary process, *i.e.* the evolution of specific genes which, in addition to determining our response to drug therapy, have a more fundamental role in protecting us from environmental insult.

The genes involved in determining pharmacogenetic variability in drug response can be broadly split into two groups. Genetic variability can occur in either the primary therapeutic targets, in the proteins which mediate their function (*e.g.* those genes which are part of cell signalling pathways) and also in genes which are specifically involved in drug uptake, metabolism and disposition. This latter group of genes, where current studies in pharmacogenetics are most advanced, represent a small tributary to the river of genes which have specifically evolved to protect multicellular organisms from the 'chemical warfare' which has been fundamental to the evolution of life on earth. These genes, whose primary role has been to protect us from ingested natural toxins in our diet, are also those involved in the uptake, metabolism and deposition of therapeutic drugs.

There are a number of factors, other than genetically determined variation, which can determine individual response to therapy. Pharmacogenetics does not encompass variability in therapeutic response, for example in antibiotic, antiparasitic or antiviral chemotherapy where lack of activity is due to genetic change in the target organism[3], although such changes do exemplify how genetic differences in general can result in enormous changes in drug sensitivity. It also does not encompass somatic mutations which may result in drug resistance in cancer chemotherapy, although the study of drug resistance has identified important target genes where allelic variants may be important. It is also important to note that drug resistance in chemotherapy is probably determined by a large number of factors, some of which do not relate to target cell biology[4]. In cancer chemotherapy, for example, dose is often limited by drug side effects. Pharmacogenetic polymorphisms may therefore also be an important determinant of the outcome of both antimicrobial and cancer chemotherapy.

Pharmacological effectiveness is critically dependent on obtaining the required concentration of drug at the desired target site for a sufficient period to generate the desired effect without inducing side effects. The mechanisms by which the drug interacts with its target, either at the cell surface or at an intracellular site are also critical determinants of drug efficacy (Fig. 1). The phenotypic consequence of polymorphic metabolism is also dependent on the therapeutic index of the drug and is most pronounced when the therapeutic index is low. It is, therefore, inevitable that allelic variants of many genes involved in drug pharmacology will exist which do not affect therapeutic response.

The evolution of pharmacogenetics

The genes which have evolved to protect us from environmental chemicals are the same as those which determine our sensitivity or resistance to the

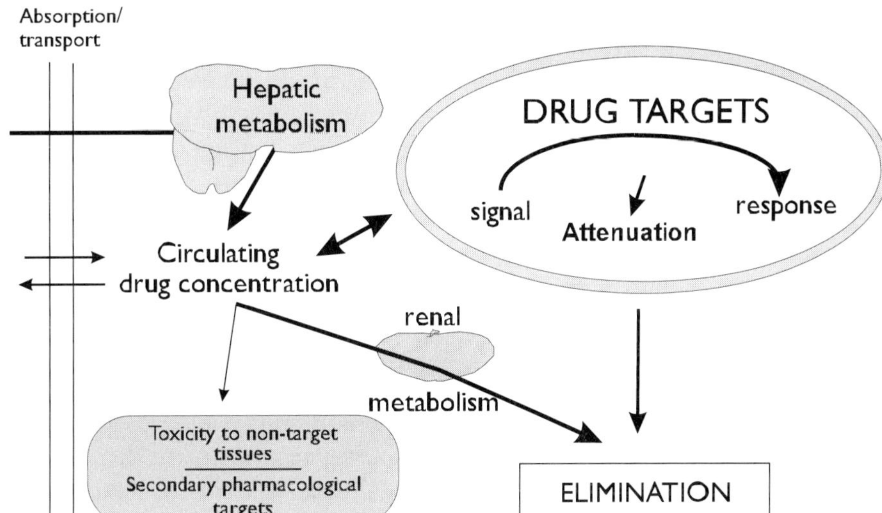

Fig. 1 Factors which can limit the therapeutic effectiveness of drugs.

effects of therapeutic drugs[5,6]. Many of these genes are polymorphic in man and there has been much debate about why this should be the case. On the one hand, it has been argued that this large degree of variability is because these genes are not essential for survival and are an evolutionary appendix which will eventually disappear. The other, more attractive, hypothesis is that the level of polymorphism reflects the heterogeneity in our global environment and the pressure on certain genes to evolve to cope with specific environmental challenges. The high level of polymorphism in the human genome may, therefore, be of critical importance for our survival. The current difficulty with this latter hypothesis, however, is that at present we have no explanation for the high prevalence of null alleles (complete loss of gene activity) in certain polymorphic genes. The selective advantage generated by the presence of null alleles still requires rationalisation.

Pharmacogenetics – a historical perspective

The birth of pharmacogenetics was brought about following a number of landmark discoveries at the end of the 19th century. As a consequence of the rapidly developing field of organic chemistry, it became apparent that most drugs were eliminated from the body with a chemical composition which differed from that which was administered, suggesting that some 'internal factor' had transformed or metabolised the parent compound. Following the pioneering work of Gregor Mendel and the establishment of the rules of heredity in 1866[7], a number of independent reports suggested that the rules of heredity also governed the biotransformation of drugs and detoxification of foreign compounds. Central to this thesis were the observations of Archibold Garrod, firstly on the physiology of human urinary pigments and secondly in the study of alcaptonuria[8]. From these early observations, Garrod in his *Inborn Errors of Metabolism*, first published in 1906[9], proposed that interindividual variation in drug response arose from 'derangement of metabolic factors'. Some years later in a classic text, R.T. Williams[10] was to propose the phase I (oxidation) and phase II (conjugation) classification of 'drug metabolising enzymes'.

Following the observations of Motulsky in 1957 that inherited defects of metabolism may explain many individual differences in the efficacy of drugs and in the occurrence of adverse drug reactions which should be distinguished from allergic reactions to drug therapy[11], 'pharmacogenetics' was first formally defined by Vogel[12] in 1959 as 'clinically important hereditary variation in response to drugs'. In 1962, the first text specifically devoted to this area *Pharmacogenetics – Heredity and the Response to Drugs* was published by Werner Kalow[13].

As the research area has developed, it has become increasingly possible to rationalise the biochemical and genetic basis for a number of seemingly unrelated idiosyncratic responses to drug therapy. Early advances in the area of pharmacogenetics came from several independent observations. Variation in serum cholinesterase levels was shown to be an inherited trait following reports that a number of individuals experienced prolonged anaesthesia following administration of the muscle relaxant suxamethonium (succinylcholine)[14]. Several allelic variants of this gene have now been characterised, and the molecular basis of the polymorphism determined[15]. These include an inactive pseudogene and a 'high-activity' form of the enzyme; the major atypical form of the enzyme arises from a single aspartic acid to glycine substitution in the anionic substrate binding site of the protein which eliminates the possibility of electrostatic interaction, abolishing substrate binding.

Similarly, reports of the development of haemolytic anaemia in significant numbers of Afro-Caribbean males following prescription of the anti-malarial drug primaquine, lead to the discovery of a hereditary defect in the gene encoding G6PD (glucose 6-phosphate dehydrogenase)[16]. G6PD, a key enzyme in the pentose phosphate pathway is now known to be highly polymorphic; the major low-activity form of the enzyme occurs as a result of a single asparagine to aspartic acid substitution and is inherited as an X-linked recessive trait[17]. More than 400 variants of the G6PD gene have now been characterised biochemically, with a variety of phenotypic consequences; more than 30 alleles have been identified which contain sequence changes in the coding sequence of the gene. In addition to the 8-aminoquinolone antimalarial drugs including primaquine, G6PD also metabolises a wide variety of drugs including chloramphenicol and the sulphonamide antibiotics.

The genetic basis of the N-acetyl transferase polymorphism was first identified in a series of elegant experiments comparing the responses of mono- and dizygotic twins to isoniazid[18] following reports of abnormal responses and the development of peripheral neuropathy in a significant number of patients when prescribed the antitubercular drug.

A further major advance in our understanding of interindividual variation in drug response was the identification of genetic variability in drug oxidation mediated by the cytochrome P450 system. Early reports of interindividual variability in drug clearance followed the prescription of antipyrene[19] and *bis*-hydroxycoumarin[20] although the first demonstration that such variability was genetically determined arose from extreme responses to the hypertensive agent debrisoquine[21] and the antiarrythmic drug sparteine[22]. Both drugs were subsequently shown to be metabolised by a single polymorphic cytochrome P450 enzyme, CYP2D6, known trivially as debrisoquine hydroxylase[23]. The major gene inactivating mutations in CYP2D6 have now been identified and a DNA based tests to identify individuals with impaired CYP2D6 metabolism developed[24,25].

The identification of the molecular basis of pharmacogenetic polymorphisms and the ability to screen individuals for the presence of specific alleles opens up the possibility of screening patients to predict drug outcome and to allow individualised drug treatment based on genetic constitution. To date, the majority of work in the field of pharmacogenetics has focused on genetic polymorphisms in human drug metabolising enzymes.

Genetic polymorphisms in drug metabolising enzymes

Nearly all lipophilic small molecules, including most drugs, which enter the body must be metabolised to more polar products before they can be excreted. This metabolic process, which is primarily catalysed by hepatic enzymes, consists of a sequence of enzymatic steps. This normally involves oxidation of the drug by the cytochrome P450-dependent monooxygenases (phase I metabolism), followed by conjugation involving sulphation, glucuronidation or acetylation (phase II metabolism). A number of enzymes including the glutathione S-transferases, N-acetyl transferases and UDP-glucuronosyl transferases are involved in catalysing these phase reactions.

A

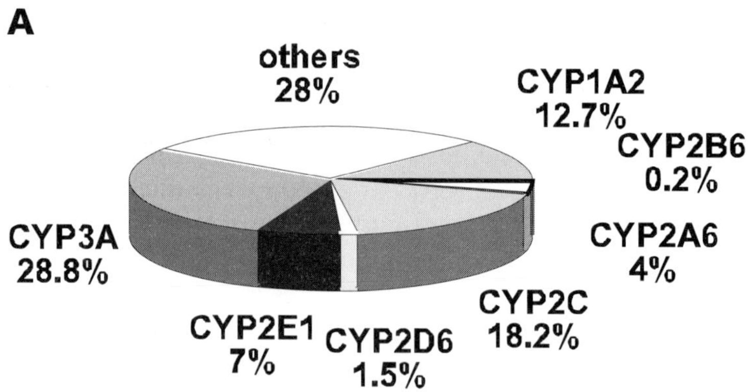

B

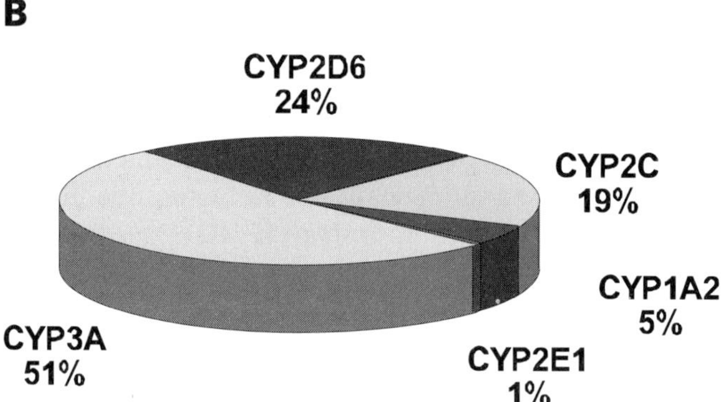

Fig. 2 (**A**) Relative abundance of P450 isoenzymes in human liver and (**B**) relative contribution of P450 enzymes to drug metabolism.

Although the enzymes involved in drug metabolism are primarily found in the liver, they are also expressed in an isozyme-specific manner in most other tissues. The vast number of environmental chemicals which they metabolise may explain why these enzymes have evolved into multigene families of proteins, with each enzyme exhibiting an unique substrate specificity[5].

Recent estimates suggest that there may be as many as 50 distinct cytochrome P450 proteins in man[26]. It is interesting to note that although many of these enzymes have the capacity to metabolise specific drugs, the majority of P450-mediated drug metabolism in man is catalysed by two enzymes, cytochrome P450 CYP3A4 and the polymorphic cytochrome P450, CYP2D6 (Fig. 2)[27].

Although cytochrome P450 CYP2D6 is only a relatively minor form in human liver (Fig. 2), one reason why it metabolises up to one quarter of all prescribed drugs could be because many of these drugs are targeted to the central nervous system (Table 1). CYP2D6 is found in the brain[28] and may, therefore, have evolved to protect cells from environmental neurotoxins[29]. In contrast, CYP3A4 is active in the metabolism of many

Table 1 CNS-acting drugs which are metabolised by one or more polymorphic P450

Drug	Metabolised by
Alprazolam	CYP3A
Amitryptiline	CYP2C19, CYP2D6, CYP1A2, CYP3A
Busiprone HCl	CYP3A
Caffeine	CYP1A2, CYP2E1
Carbemazepine	CYP3A, CYP2C8
Chlorpromazine	CYP1A1, CYP2B6, **CYP2D6**, CYP3A
Clomipramine	CYP2C19, **CYP2D6**, CYP1A2, CYP3A
Clozapine	CYP1A2, **CYP2D6**, CYP2C9, CYP2C8, CYP2C19, CYP2E1, CYP3A
Diazepam	CYP2C18, CYP2C19, CYP3A
Desipramine	**CYP2D6**
Fluoxetine	CYP2C19, **CYP2D6**, CYP3A, CYP2C9, CYP2E1
Fluphenazine	**CYP2D6**
Fluvoxamine	CYP1A2, CYP2C19, CYP3A, **CYP2D6**, CYP2C9
Haloperidol	**CYP2D6**, CYP3A, CYP1A2, CYP3A
Imipramine	**CYP2D6,** CYP1A2, CYP3A4, CYP2C19
Levopromazine	**CYP2D6**
Mianserin	CYP1A2, CYP2B6, CYP2C19, **CYP2D6**, CYP3A
Midazolam	CYP3A, CYP4B1
Moclobemide	CYP2C19
Nortryptiline	**CYP2D6**
Paroxetine	**CYP2D6**
Perphenazine	**CYP2D6**
Phenytoin	CYP2C19, CYP2C9, CYP3A
Risperidone	**CYP2D6**
Sertraline	CYP2C9, **CYP2D6**, CYP3A
Temazepam	CYP2C8, CYP2C19, CYP3A
Thioridazine	**CYP2D6**
Tranylcypromine	CYP2A6, CYP2C19
Trazodone	CYP3A
Triazolam	CYP3A
Trifluperidol	**CYP2D6**
Trimipramine	**CYP2D6**
Zuclopenthixol	**CYP2D6**

Drugs metabolised by **CYP2D6** are highlighted in **bold**

naturally occurring antibiotics which may rationalise its central role in drug metabolism. It is important to note, however, that although CYP3A4 and CYP2D6 catalyse the majority of drug oxidations in man, other P450 enzymes can play a pivotal role in the deposition of certain drugs. The cytochrome P450 system is our primary defence against ingested chemicals and determines the biological half-life of the vast majority of drugs which are in current clinical use.

There are also many examples of prodrugs which are converted to their active form as a consequence of cytochrome P450-mediated oxidation reactions. These include the activation of the anticancer drug cyclophosphamide to its principal cytotoxic metabolite[30] and the conversion of codeine to the analgesic morphine[31].

As an unfortunate consequence of the nature of the reactions catalysed by P450 monooxygenases, however, P450-mediated metabolism can also convert a therapeutically-useful drug into a toxic compound which produces unwanted side effects. This is exemplified by the metabolism of paracetamol to NABQI (N-acetyl-*p*-benzoquinone immine), the metabolite responsible for its hepatotoxicity[32]. It is also the major pathway of toxicity and carcinogenicity for the pulmonary carcinogens found in cigarette smoke such as benzo(a)pyrene and dimethylbenz-anthracene. These undesirable metabolic activation reactions can be a major reason why many potential new drugs fail in development and can account for the incidence of many drug side effects.

Individuality in the expression of P450 isozymes can, therefore, have a variety of therapeutic consequences including: (i) no therapeutic effect due to ultra-rapid drug metabolism and clearance; (ii) lack of prodrug activation; (iii) toxic side effects due to drug accumulation as a consequence of impaired metabolism; or (iv) metabolic activation to toxic products[38]. In cancer chemotherapy, individuality in treatment response could also be due to variations in the levels of drug metabolising enzymes in the target tumour tissues.

It is well known that modulation of cytochrome P450 function can have a profound effect on drug pharmacology. This is not only based on genetic studies, but also from our knowledge of adverse drug/drug reactions which arise when one drug (or compound) inhibits the P450 mediated metabolism of an other taken concomitantly. One interesting example of this is the capacity of the furanocoumarin, bergamottin, and the flavinoid, naringenin, both of which are found in grapefruit juice, to inhibit CYP3A4 activity[33]. This can result in adverse reactions to drugs such as the HIV protease inhibitor saquinivir, lovastatin, a cholesterol-lowering drug and the immunosuppressive drug cyclosporin[34,35].

The cytochrome P450 system is an adaptive response to environmental challenge and, as a result, exposure to a chemical can often result in the induction of a P450 isozyme active in its metabolism. This mechanism

Table 2 The major polymorphic drug-metabolising P450s

Enzyme	Chromosomal localisation	No. of alleles	Major alleles
CYP1A1	15q 22-qter	4	
CYP1B1	2p22-21	6	
CYP2A6	19q13.1-13.3	3	Consensus 17% v1, T→A Exon3, 78% v2, C→A Exon 3, 7%
CYP2C9	10q24.1-24.3	3	Consensus 79% CYP2C9*2, C→T Exon 3, 12.5% CYP2C9*3, A→C Exon 7, 8.5%
CYP2C19	10q24.1-24.3	10	Consensus 75–79% CYP2C19*2, G→A Exon 5, 22-29% CYP2C19*3, G→A Exon 4
CYP2D6	22q11.2-qter	>18	Consensus 65–70% CYP2D6*3, del A_{2637}, approx. 4% CYP2D6*4, $G_{1934}A$, 10–20% CYP2D6*5, gene deletion, approx. 4%
CYP2E1	10	3	

was first proposed by Conney in 1967, who observed that several drugs, including phenylbutazone, were capable of enhancing their own metabolism[36]. Such effects result in large individual variability in hepatic cytochrome P450 isozyme expression[37]. Genetically-determined variability in drug response is, therefore, often superimposed on a significant level of metabolic variability and can often be masked, particularly if individual allelic variants only differ slightly in their kinetic properties.

In spite of these potentially confounding factors, genetic polymorphism in the cytochrome P450 system can have marked effects on therapeutic outcome. A number of cytochrome P450 genes have now been shown to be polymorphic – these are summarised in Table 2. By far the most extensively studied of these genes, however, is CYP2D6.

Approximately 6% of the Caucasian population carry two null alleles at the CYP2D6 gene locus. In the UK alone, therefore, approximately 3.6 million people do not express this enzyme and, as a consequence, have a compromised ability to metabolise the wide range of therapeutic drugs which are CYP2D6 substrates (Table 3). A variety of phenotypic consequences have been associated with the CYP2D6 null genotype (Table 4)[38]. These include rare lethal adverse drug reactions, for example following administration of the antianginal drug, perhexiline, or lack of therapeutic efficacy due to either ultrarapid metabolism or lack of prodrug activation.

In addition to the well-characterised null alleles of CYP2D6, there are several additional alleles containing single amino acid changes[39], certain of which have also been associated with altered phenotype. Many CYP2D6 alleles, however, do not yet have a clearly defined phenotype.

Table 3 CYP2D6 drug substrates

Alprenolol	Amiflavine
Amiodorone	Amitryptiline
Apigenin	Budesonide
Bufuralol	Bupranolol
Chloral hydrate	Clomipramine
Clonidine	Clotrimazole
Clozapine	Codeine
Cyclobenzaprine	Desipramine
Dexfenfluramine	Dextromethorphan
Dibucaine	Dihydroergotamine
Dolasetron	Doxorubicin
Encainide	Ethinylestradiol
Ethylmorphine	Fenoterol
Flecainide	Fluvoxamine maleate
Fluoxetine	Formoterol
Guanoxan	Haloperidol
4-hydroxy amphetamine	Imipramine
Indoramine	Ketoconazole
Laudanosine	Levomepromazine
Loratadine	MDMA (ecstacy)
Mefloquine	Methoxamine HCl
Methoxyphenamine	Methoxypsoralen
Methysergide HCl	Metoclopramide
Metoprolol	Minaprine
Moclobemide	MPTP
Mexiletine	Nicergoline
Nimodipine	Nitrendipine
Nortriptyline	Olanzapine
Ondansetron	Oxprenolol
Paroxetine	Perhexiline
Perphenazine	Phenformin
Phenylpropanolamine	Procainamide
Promethazine	N-propylajmaline
Propafenone	Propranolol
Pyrimethamine	Quercitin
Rifampicin	Ritonavir
Roxithromycin	Serotonin
Sparteine	Sulfasalazine
Tacrine	Tamoxifen
Thioridazine	Timolol
Tomoxetine	Tranylcypromine
Tropisetron	Zuclopenthixol

Inheritance of these rare CYP2D6 alleles does not at present therefore clearly define a susceptible group and their usefulness in predicting therapeutic response remains uncertain.

A further interesting polymorphism at the CYP2D6 gene locus was identified from the observation that certain individuals, 'ultra-rapid' metabolisers, did not respond, or had only a very limited response, to therapy with drugs which were CYP2D6 substrates. This phenomenon

Table 4 CYP2D6 – implications for poor metabolisers

Decreased elimination - accumulation of parent compound resulting in toxicity		
	β-Blockers	Metoprolol Timolol Bufuralol
	Anti-arrythmics	Sparteine Mexiletine Flecainide
	Tricyclic Antidepressants	Nortryptiline Clomipramine
	Neuroleptics	Perphenazine Zuclopenthixol
Decreased prodrug activation		
	Encainide	(active metabolite: O-desmethyl encainide)
	Codeine	(active metabolite: morphine)
Decreased elimination of active metabolite		
	Imipramine	(active metabolite: desipramine)
Decreased elimination of parent compound and active metabolite		
	Amitryptyline	(active metabolite: nortryptyline)

was subsequently shown to be due to amplification of the CYP2D6 gene, with certain individuals inheriting 2, 3 or as many as 13 tandemly arranged copies of CYP2D6[40]. Once identified, such individuals require significantly elevated doses of CYP2D6 drugs in order to achieve an appreciable clinical effect[41].

A number of further P450 genes are also genetically polymorphic (Table 2). Certain of these, most notably CYP2C19, also contain gene-inactivating or null alleles, with obvious phenotypic consequences. CYP2C19 metabolises a smaller range of drugs, but some CYP2C19 drug substrates including diazepam, omeprazole and proguanil are commonly prescribed[42]. For a number of other genes, *e.g.* CYP2C9, alleles containing only single amino acid differences clearly also generate a clinical phenotype[42,43]. This is exemplified by the observations of Rettie *et al.*, who reported a detailed case-history of a patient, subsequently shown to be homozygous for an infrequent CYP2C9 low-activity allele, who developed life-threatening warfarin toxicity during treatment for a myocardial infarction[44].

One of the most recently isolated human P450s, CYP1B1, is also polymorphic[45], although the phenotypic consequences of allelic variation in this gene are not yet fully understood. This gene is particularly interesting as, in addition to its role in drug and xenobiotic metabolism, CYP1B1 has an endogenous role in oestrogen metabolism, catalysing the 4-hydroxylation of oestrogen to oestradiol[46]. Intriguingly, mutations in CYP1B1 have also been associated with the development of primary congenital glaucoma[47].

A number of phase II enzymes, including those within the glutathione S-transferase, N-acetyl transferase, UDP-glucuronosyl transferase and sulphotransferase gene families, are also polymorphic. The glutathione S-transferase GSTM1 and GSTT1 genes are frequently deleted and a further gene, GSTP1, has allelic variants containing single amino acid substitutions which have been shown to influence both protein stability and substrate specificity[48,49]. Recent experiments in a line of transgenic mice where the Gstp genes have been deleted from the mouse genome demonstrated a significantly higher rate of chemically-induced skin papillomas, when compared to their wild-type littermates[50].

The genetic basis for the acetylation polymorphism is also now well understood and is due to the presence of multiple alleles at the NAT2 gene locus[51]. Polymorphic variation in this enzyme, responsible for the metabolism of a wide range of commonly prescribed drugs including the antitubercular drug, isoniazid, the stimulant, caffeine, and a range of heterocyclic amine carcinogens, results in a trimodal distribution of acetylation phenotypes. Although the direct effects of polymorphisms in phase II enzymes are often less pronounced than genetically-determined variation in the expression of specific P450 isozymes, the effects of inherited variation in both phase I and phase II metabolism can often be synergistic.

Genetically determined differences in sensitivity to alcohol were first reported some 20 years ago[52]. A number of possible mechanisms were proposed to account for the different phenotypes observed following alcohol ingestion, including genetically determined differences in alcohol metabolism. In addition to metabolism by the polymorphic P450, CYP2E1, ethanol is also metabolised by a number of other polymorphic pathways. Ethanol is first converted to acetaldehyde by the dimeric cytosolic enzyme alcohol dehydrogenase (ADH), a phenotypic variant of which is known to have significantly increased catalytic activity. The molecular basis for this increase in catalytic activity has now been determined and has been attributed to a single arginine to histidine substitution in the β-subunit of the ADH2 gene[53]. Acetaldehyde is then further metabolised by another polymorphic cytosolic enzyme, aldehyde dehydrogenase (ALDH), which is known to have significantly reduced activity in certain populations[54]. Inheritance of the various allelic forms of these polymorphic enzymes has been associated with differential susceptibility to the effects of alcohol in populations of differing ethnic origin[55]. Although the data is much less convincing, genetically determined differences in ethanol metabolism have also been associated with susceptibility to alcoholism and alcoholic liver disease.

Thiopurine methyl transferase (TPMT) catalyses the conjugation of the methyl group of S-adenosyl methionine to aromatic and heterocyclic sulphhydryl substrates. These include the anti-leukaemic agents 6-mercaptopurine and 6-thioguanine and azathioprine, an immunosuppressive agent used to prevent rejection in transplant surgery and in the treatment of a number of autoimmune diseases including rheumatoid arthritis and multiple sclerosis[56]. TPMT exhibits a trimodal distribution of phenotypes which has recently been rationalised by the identification of a number of allelic variants at the TPMT gene locus in addition to the presence of an inactive pseudogene[57]. Several TPMT alleles have now been identified which contain gene-inactivating mutations; inheritance of these forms of the enzyme has been shown to be responsible for the failure of drug therapy for example, in the treatment of childhood leukaemia with agents which are TPMT substrates. TPMT is, therefore, an ideal candidate gene where DNA-based genotyping tests may be implemented to routinely screen susceptible patient populations prior to drug treatment.

Human 'drug metabolising enzymes' have obviously not evolved exclusively to metabolise the drugs which are prescribed today. As an unfortunate consequence of the almost infinitely wide substrate specificity of these enzymes, a vast number of toxic, mutagenic and/or carcinogenic substrates have also been identified. Allelic variation in these enzymes is therefore likely to be a significant determinant of how we respond to environmental insult and has been implicated in individual susceptibility to a wide range of diseases including Parkinson's disease and various types of cancer. The strength of many of these associations, however, requires further substantiation.

Ethnic differences in drug metabolism

There are marked ethnic differences in the distribution of allelic variants of many drug metabolising enzymes. It is of obvious importance that such ethnic differences are taken into account during routine pharmacogenetic screening in drug discovery and, perhaps more importantly, when drugs which are subject to polymorphic metabolism are prescribed to patients from different ethnic backgrounds. Alleles which are not present or are present at only very low frequencies in the Caucasian population may be significant determinants of drug response in other populations, making specific ethnic groups much more sensitive or refractory to a particular drug regime. With the increasing level of ethnic heterogeneity within the human population, it is likely that any pharmacogenetic screening will involve screening each patient for all the known alleles of a particular polymorphic gene, irrespective of their ethnic distribution, which are known to have phenotypic consequences.

In view of the current technological advances in high-throughput genotyping this should not pose a significant problem.

Again taking CYP2D6 as an example, the PM genotype is found in approximately 6% of the Caucasian population but in only 2% of American Blacks and in less than 1% of Orientals[58]. The genetic basis for this difference has been investigated in Oriental populations and has been attributed to the relatively low abundance of the common 'Caucasian' CYP2D6 null alleles in Oriental populations. In contrast, however, further alleles of CYP2D6 which occur very infrequently in Western populations are relatively abundant in Orientals. These 'Oriental' alleles do not contain gene-inactivating mutations but do have reduced activity[59]. As a consequence of the relatively high frequency of these alleles in Orientals, anti-depressant drugs which are substrates for CYP2D6 are routinely prescribed at lower doses in Oriental compared to Caucasian populations[58]. A similar situation exists for CYP2C19 substrates; the CYP2C19 null phenotype affects only 3–5% of Caucasian populations, but more than 20% of Chinese[60]. This is reflected in the significant reduction in dosing regimens in Orientals for drugs including diazepam which are CYP2C19 substrates[61].

There is also considerable inter-ethnic heterogeneity in the frequency of the 'ultra-rapid' metaboliser gene amplification alleles of CYP2D6 which occur at frequencies of 20%, 7% and 1.5% in Ethiopian, Spanish and Scandinavian populations, respectively [62].

Significant ethnic differences are also seen a variety of other drug metabolising enzymes including NAT2[63]. As a number of these alleles have now been variously associated with disease susceptibility, there are obvious implications for differential disease susceptibilities in ethnically distinct populations.

Recent developments in pharmacogenetics

The application of genetic testing in the health service remains a difficult and emotive issue. Clear benefits can already be obtained for individuals who may carry genes predisposing them diseases with a high degree of familial inheritance, such as certain neurodegenerative disorders and family cancer syndromes. However, other applications of genetics to the health service where, for example, relatively low penetrance alleles may be makers of a particular phenotype or disease may be more difficult to implement.

Pharmacogenetics does identify an area where genetic screening as applied to drug use, may provide significant benefits to the patient and could also be of enormous economic importance. This leads to the somewhat futuristic view of tailoring drug use to the unique genetic makeup of each individual patient, where a general practitioner would generate a pharmacogenetic profile of each patient and prescribe drugs accordingly. The tremendous recent advances in genotyping technology means that obtaining such genetic information will certainly be feasible.

There is currently great interest in the pharmaceutical and biotechnology industries in increasing both our knowledge and the application of pharmacogenetics. The obvious area of current study is to screen for polymorphisms in genes involved in drug uptake and efflux, further genes involved in drug metabolism and also in the genes which regulate the expression of these primary drug targets such as cell surface or nuclear receptors. This targeted approach will undoubtedly uncover novel functional polymorphisms which influence drug response.

The other important area of rapid growth in pharmacogenetic testing is in the pharmaceutical industry, where an increasing number of companies are genotyping their clinical trial populations, in part because knowledge of genetic variability in drug response is becoming an increasingly important component of the drug registration process. It is obviously also prudent to exclude potentially susceptible individuals from phase I dose escalation trials. Genetic factors influencing the way a drug is metabolised can, therefore, have significant commercial implications. In a simplistic sense, the ability to screen new lead compounds *in vitro* in order to identify genes involved in their uptake and metabolism may allow the prediction of whether a polymorphic enzyme may be a important factor in drug deposition. With this knowledge, it may be possible in some cases to redesign the drug so that these pathways are avoided, while retaining pharmacological activity. The merit of such an approach remains unclear, however, as it can significantly increase development time and it is not clear whether further polymorphism at other unrecognised sites will also contribute to individuality in drug response.

At present, the assumption that demonstrating a polymorphic route of metabolism for a particular lead compound will compromise its use or application remains at best equivocal and is a difficult clinical and commercial decision. In any event, polymorphism in drug targets will almost certainly be unavoidable. Ultimately, we are all genetically different and it will be impossible to design out all of the genetically-determined variability in drug response. If we are in future to move away from the current empiricism associated with drug prescription and avoid the interindividual variations in drug response and side effects which currently compromise the use of many drugs, the application of pharmacogenetic screening to identify susceptible individuals will have a very important role to play.

Recent advances in both mutation detection and genotyping technologies have greatly reduced the technical limitations of screening large populations to generate individual 'genetic fingerprints', which could be used as the basis for rational prescribing. Many of the analytical techniques are now fully automated and require only very small 'pin-prick' quantities of blood. The use of peripheral blood and other readily accessible tissues such as hair follicles and buccal scrapings has become routine in DNA-based genotyping analysis. It may also prove possible to use these tissues as 'surrogate markers', in the development of minimally invasive methods to produce individual 'metabolic profiles' in order to assess individual drug metabolising capabilities.

Where next?

There are obvious ways forward in the development and application of pharmacogenetics. A more comprehensive knowledge of polymorphisms at specific candidate gene loci needs to be acquired and the pharmacological significance of these polymorphisms needs to be determined in prospective studies. Such studies will allow us to determine the degree to which individual sensitivity to specific classes of drug is genetically determined. Although the potential importance of genetic variability in drug response is generally acknowledged in academic circles, the pharmaceutical industry and the drug regulatory authorities, this is not yet the case in general practice and, indeed, in many clinical pharmacology departments. A greater awareness of this is urgently needed. DNA-based methods have been developed for high throughput routine screening and have already been clinically applied. Many of the drugs metabolised by CYP2D6, for example, are CNS-active agents which have narrow therapeutic indices and drug over-treatment and accumulation can give rise to symptoms similar to those of the disease itself. Genetic variability can be a major cause of variations in drug plasma concentration. The difficulties in distinguishing between lack of therapeutic response relative to the effects of drug overdosing becomes increasingly difficult to distinguish if the patient being treated has a psychiatric illness or is suffering for acute depression.

Key points for clinical practice

- Adverse drug reactions are a significant clinical problem and a major cause of death.

- Pharmacogenetics is defined as the genetic basis of individuality in response to drugs.
- Pharmacogenetic variability can explain lack of therapeutic efficacy and can also rationalise idiosyncratic adverse drug reactions.
- Individuality in drug response can result from genetically-determined variation in the absorption, disposition, metabolism and excretion of drugs and also from polymorphisms in target sites.
- The best characterised pharmacogenetic polymorphisms are those within the cytochrome P450 family.
- CYP2D6, debrisoquine hydroxylase, is by far the most extensively characterised of the polymorphic P450s. CYP2D6 metabolises up to 25% of all commonly prescribed drugs and is inactive in 6% of the Caucasian population.
- Inheritance of specific allelic variants of many of the polymorphic drug metabolising enzymes may be markers for disease susceptibility.
- A detailed knowledge of the genetic basis of individual drug response is of significant clinical and economic importance and may provide the basis for rational drug prescription.

Acknowledgement

GS thanks the Ministry of Agriculture, Fisheries and Food (Project FS1732) for financial support.

References

1 Krantz JC, Carr CJ. eds. The Pharmacologic Principles of Medical Practice. Baltimore: Williams & Wilkins, 1954
2 Lazarou J, Pomeranz BH, Corey PN. Incidence of adverse drug reactions in hospitalized patients: a meta-analysis of prospective studies. JAMA 1998; **279**: 1200–5
3 Hayes JD, Wolf CR. Molecular mechanisms of drug resistance. Biochem J 1990; **272**, 281–95
4 Hayes JD, Wolf CR (eds). Molecular Genetics of Drug Resistance. London: Harwood Academic, 1997
5 Wolf CR. Cytochrome P-450s: polymorphic multigene families involved in carcinogen activation. Trends Genet 1986; **2**, 209–14
6 Gonzalez FJ, Nebert DW. Evolution of the P450 gene superfamily: animal-plant 'warfare', molecular drive and human genetic differences in drug oxidation. Trends Genet 1990; **6**, 182–6
7 Mendel JG. Versuche über Pflanzen-Hybride. Verhandlungen des naturforschenden Vereines in Brünn 1866; **4**

8 Garrod AE. The incidence of alcaptonuria: a study in chemical individuality. Lancet 1902; ii: 1616–20

9 Garrod AE. Inborn Errors of Metabolism. New York: Oxford University Press, 1909.

10 Williams RJ. Biochemical Individuality. New York: Wiley, 1956

11 Motulsky AG. Drug reactions, enzymes and biochemical genetics. JAMA 1957; **165**: 835–7

12 Vogel F. Moderne probleme der Humangenetik. Ergeb Inn Med Kinderheild 1959; **12**: 52–125

13 Kalow W. Pharmacogenetics – Heredity and the Responses to Drugs. Philadelphia: WB Saunders, 1962

14 Forbat A, Lond MB, Lehmann H, Silk E. Prolonged apnoea following injection of succinylcholine. Lancet 1953; ii: 1067–8

15 LaDu BN, Bartels CF, Nogueira CP et al. Phenotypical and molecular biological analysis of human butyrylcholinesterase variants. Clin Biochem 1990; **23**: 423–31

16 Hockwald RS, Arnold J, Clayman CB, Alving AS. Toxicity of primaquine to Negroes. JAMA 1952; **149**: 1568–70

17 Beutler E. The molecular biology of G6PD variants and other red cell enzyme defects. Annu Rev Med 1992; **43**: 47–59

18 Bönicke R, Lisboa BP. Über die Erbbedingtheit der intraindividuellen Konstanz der Isoniazidausscheidung beim Menschem (Untersuchungen an eineiigen und zweieiigen Zwillingen). Naturwissenschaften 1956; **44**: 314

19 Brodie BB, Axelrod J. The fate of antipyrene in man. J Pharmacol Exp Ther 1950; **98**: 97–104

20 Weiner M, Shapiro S, Axelrod J, Cooper JR, Brodie BB. The physiological disposition of dicoumarol in man. J Pharmacol Exp Ther 1950; **99**: 409–20

21 Mahgoub BA, Idle JR, Dring LG et al. Polymorphic oxidation of debrisoquine in man. Lancet 1977; i:584–6

22 Eichelbaum M, Spannbrucker N, Steincke B et al. Defective N-oxidation of sparteine in man: a new pharmacological defect. Eur J Clin Pharmacol 1979; **16**: 183–7

23 Gonzalez FJ, Skoda RC, Kimura S et al. Characterisation of the common genetic defect in humans deficient in debrisoquine metabolism. Nature 1988; **331**: 442–6

24 Gough AC, Miles JS, Spurr NK et al. Identification of the primary gene defect at the cytochrome P450 CYP2D gene locus. Nature 1990; **347**: 773–6

25 Heim M, Meyer UA. Genotyping of poor metabolisers by allele-specific PCR amplification. Lancet 1990; **336**: 529–32

26 Nelson DR, Koymans L, Kamataki T et al. P450 superfamily: update on new sequences, gene mapping, accession numbers and nomenclature. Pharmacogenetics 1996; **6**: 1–42

27 Shimada T, Yamazaki H, Mimura M et al. Inter-individual variations in human liver cytochrome P-450 enzymes involved in the oxidation of drugs, carcinogens and toxic chemicals: studies with liver microsomes of 30 Japanese and 30 Caucasians. J Pharmacol Exp Ther 1994; **270**: 414–23

28 Gilham DE, Cairns W, Paine MJI et al. Metabolism of MPTP by cytochrome P4502D6 and the demonstration of 2D6 mRNA in human foetal and adult brain by in situ hybridisation. Xenobiotica 1997; **27**: 111–25

29 Smith CAD, Gough AC, Leigh PN et al. Debrisoquine hydroxylase gene polymorphism and susceptibility to Parkinson's disease. Lancet 1992; **339**: 1365–72

30 Hadidi AHFA, Coulter CEA, Idle JR. Phenotypically deficient urinary elimination of carboxyphosphamide after cyclophosphamide administration to cancer patients. Cancer Res 1988; **48**: 5167–71

31 Yue QY, Svensson JO, Alm C, Sjoqvist F, Sawe J. Codeine O-demethylation co-seggregates with polymorphic debrisoquine hydroxylation. Br J Clin Pharmacol 1989; **28**: 639–45

32 Raucy JL, Lasker JM, Leiber CS, Black M. Acetaminophen activation by human liver cytochromes P-450IIE1 and P-450IA2. Arch Biochem Biophys 1989; **271**: 270–83

33 Fuhr U. Drug interactions with grapefruit juice. Extent, probable mechanism and clinical relevance. Drug Safety 1998; **18**: 251–72

34 Kupferschmidt HH, Fattinger KE, Ha HR, Follath F, Krahenbuhl S. Grapefruit juice enhances

the bioavailability of the HIV protease inhibitor saquinivir in man. Br J Clin Pharmacol 1998; **45**: 355–9

35 Ameer B, Weintraub RA. Drug interactions with grapefruit juice. Clin Pharmacokinet 1997; **33**: 103-21

36 Conney AH. Pharmacological implications of microsomal enzyme induction. Pharmacol Rev 1967; **19**: 317–66

37 Forrester LM, Henderson CJ, Glancey MG et al. Relative expression of cytochrome P450 isozymes in human liver and association with the metabolism of drugs and xenobiotics. Biochem J 1992; **281**: 359–68

38 Tucker GT. Clinical implications of genetic polymorphisms of drug metabolism. J Pharmacol 1994; **46**: 417–24

39 Daly AK, Brockmuller J, Broly F et al. Nomenclature for human CYP2D6 alleles. Pharmacogenetics 1996; **6**; 193–201

40 Johansson I, Lundquist E, Bertilsson L et al. Inherited amplification of an active gene in the cytochrome P450 CYP2D locus as a cause of ultrarapid metabolism of debrisoquine. Proc Natl Acad Sci USA 1993; **90**: 11825–9

41 Bertilsson L, Dahl ML, Sjoquist F et al. Molecular basis for rational prescribing in ultra-rapid hydroxylators of debrisoquine. Lancet 1993; **341**: 63

42 Miners JO, Birkett DJ. Cytochrome P4502C9: an enzyme of major importance in human drug metabolism. Br J Clin Pharmacol 1998; **45**: 525–38

43 Stubbins MJ, Harries LW, Smith G, Tarbit MH, Wolf CR. Genetic analysis of the human cytochrome P450 CYP2C9 locus. Parmacogenetics 1996; **6**: 429–39

44 Steward DJ, Haining RL, Henne KR et al. Genetic association between sensitivity to warfarin and expression of CYP2C9*3. Pharmacogenetics 1997; **7**: 361–7

45 Bailey LR, Roodi N, Dupont WD, Parl FF. Association of cytochrome P450 1B1 (CYP1B1) polymorphism with steroid receptor status in breast cancer. Cancer Res 1998; **58**: 5038–41

46 Hayes CL, Spink DC, Spink BL et al. 17β-Estradiol hydroxylation catalysed by cytochrome P4501B1. Proc Natl Acad Sci USA 1997; **93**: 9776–81

47 Stoilov I, Akarsu AN, Alozie I et al. Sequence analysis and homology modelling suggest that primary congenital glaucoma on 2p21 results from mutations disrupting either the hinge region or the conserved core structures of cytochrome P4501B1. Am J Hum Genet 1998; **62**: 573–84

48 Zimniak P, Nanduri B, Pilula S et al. Naturally occurring human glutathione S-transferase GSTP1.1 isoforms with isoleucine and valine at position 104 differ in enzymatic properties. Eur J Biochem 1994; **244**: 893–9

49 Harries LW, Stubbins MJ, Forman D, Howard GCW, Wolf CR. Identification of genetic polymorphisms at the glutathione S-transferase Pi locus and association with susceptibility to bladder, testicular and prostate cancer. Carcinogenesis 1997; **18**: 641–4

50 Henderson CJ, Smith AG, Ure J, Brown K, Bacon EJ, Wolf CR. Increased skin tumorigenesis in mice lacking pi class glutathione S-transferases. Proc Natl Acad Sci USA 1998; **95**, 5275–80

51 Blum M, Demierre A, Grant DM, Heim M, Meyer UA. Molecular mechanism of slow acetylation of drugs and carcinogens in humans. Proc Natl Acad Sci USA 1991; **88**: 5237–41

52 Vessell ES, Page JG, Passanti GT. Genetic and environmental factors affecting ethanol metabolism in man. Clin Pharmacol Ther 1971; **12**: 192–201

53 Bühler R, Hempel J, von Wartburg JP, Jörnvall H. Human alcohol dehydrogenase: structural differences between the b and g subunits suggest parallel duplication in isozyme evolutions and predominant expression of separate gene descendants in livers of different mammals. Proc Natl Acad Sci USA 1984; **81**: 6320–4

54 Goedde HW, Argawal DP. Pharmacogenetics of aldehyde dehydrogenase (ALDH). Pharmacol Ther 1990; **45**: 345–71

55 Mizoi Y, Adachi J, Fukanaga T et al. Individual and ethnic differences in ethanol elimination. Alcohol Alcohol 1987; **1**: 389–94

56 Lennard L. The clinical pharmacology of 6-mercaptopurine. Eur J Clin Pharmacol 1992; **43**: 329–39

57 Krynetski EY, Scheutz JD, Galpin AM, Pui CH, Relling MV, Evans WE. A single point mutation leads to loss of catalytic activity in human thiopurine S-methyl transferase. Proc Natl Acad Sci USA 1995; **92**: 949–53

58 Kalow W, Bertilsson L. Interethnic factors affecting drug response. Adv Drug Res 1994; **25**: 1–53

59 Lou YC, Ying L, Bertilsson L, Sjoqvist F. Low frequency of slow debrisoquine hydroxylation in a native Chinese population. Lancet 1987; **ii**: 852–3

60 Wilkinson GR, Guengerich FP, Branch RA. Genetic polymorphism of S-mephenytoin hydroxylation. Pharmacol Ther 1989; **43**: 53–76

61 Wedlund PJ, Aslanian WS, Jacqz E et al. Phenotypic differences in mephenytoin pharmacokinetics in normal subjects. J Pharmacol Exp Ther 1985; **234**: 662–9

62 Aklillu E, Persson I, Bertilsson L, Rodriques F, Ingelman-Sundberg M. Frequency distribution of ultra-rapid metabolisers of debrisoquine in a black African population carrying duplicated and multiduplicated CYP2D6 alleles. J Pharmacol Exp Ther 1996; **278**: 441–6

63 Vatsis KP, Weber WW. *Acetyltransferases. Biotransformation Comprehensive Toxicology.* Oxford: Elsevier, 1997

Pathogen virulence genes – implications for vaccines and drug therapy

Christoph Tang* and **David Holden**[†]

**Department of Paediatrics, John Radcliffe Hospital, Oxford, UK and †Department of Infectious Diseases, Imperial College of Science, Technology and Medicine, Hammersmith Hospital, London, UK*

The emergence and spread of bacteria resistant to antimicrobial drugs is a major public health problem with a growing number of infections becoming virtually untreatable. There is a need to develop interventions both to prevent and to treat diseases caused by multi-resistant microbes. We review some recently developed methods (including whole genome nucleotide sequencing projects) to study bacterial pathogenesis, and discuss how knowledge gained from understanding the molecular mechanisms of disease can be applied to combat the threat of infectious diseases.

Our ability to treat and prevent bacterial infections is becoming increasingly inadequate. Over the past decade, bacterial pathogens have continued to be leading causes of mortality in developing countries, and have re-emerged as important public health problems in developed countries. This represents a significant change from the position in the 1970s and early 1980s. Until that time, bacterial infections appeared to be waning in the face of improvements in living conditions and the implementation of vaccination programmes. Furthermore, there were sufficient antimicrobial drugs available to withhold classes of compounds for life-threatening infections. However, no new class of antimicrobial has become available in the past 20 years, and the therapeutic reserve has now run dry. We are currently faced with the prospect of untreatable systemic bacterial infections caused by important human pathogens such as *Staphylococcus aureus* and *Enterococcus faecium*[1,2]. Therefore there is an urgent need to identify new therapies and vaccines to treat and/or prevent bacterial infections.

The global importance of the emergence and spread of antimicrobial resistance has been recognised by the US Centers for Disease Control and the World Health Organization[3], resulting in the International Surveillance Program for Emerging Antimicrobial Resistance in Hospitals (INSPEAR). This collaborative venture is designed to monitor antimicrobial susceptibility from 1999 in hospitals across 25 countries.

Correspondence to: Prof. David Holden, Department of Infectious Diseases, Imperial College of Science, Technology and Medicine, Hammersmith Hospital, Du Cane Road, London W12 0NN, UK

Table 1 Bacterial pathogens causing increased concern

Problem	Example
Multiply drug resistant	*Enterococcus spp.* *Mycobacterium tuberculosis* *Salmonella typhi* *Salmonella typhimurium* *Staphylococcus aureus* *Streptococcus pneumoniae* *Yersinia pestis*
Newly identified pathogens	*Borrelia burgdorferi* *Ehrlichosis* *Escherichia coli* O157 *Legionella pneumophila*
Nosocomial pathogens	*Acinetobacter baumannii* *Clostridium difficle* Coagulase-negative *Staphylococci* *Enterococcus* spp. *Pseudomonas aeruginosa* *Serratia marcescens*

The bacteria responsible for the current precarious situation can be broadly classified into three groups (Table 1). Firstly, several common pathogens have become multiply drug resistant, in part through the increased and inappropriate medical and agricultural use of anti-microbials. Secondly, 'new' pathogens have been described that result from exposure of individuals through travel or changes in lifestyle (*e.g.* animal husbandry, food processing and urbanisation). Furthermore, increasing medical intervention has rendered people susceptible to a wide range of organisms by reducing physical (*e.g.* cannulation and catheterisation) and immunological barriers to infection.

In the last few years, our understanding of the molecular basis of virulence of certain well-studied bacterial pathogens has increased dramatically. This has resulted from the application of recombinant DNA technology and cell biology to investigate bacterial infections, and the development of genetic techniques for identifying virulence genes. More recently, information about bacterial pathogens has become available from bacterial genome sequencing projects. Many who work in the field hope this understanding will be exploited to develop new interventions against bacterial infections. We will discuss what has been learnt from recent progress in our knowledge of the biology of bacterial pathogens, and how this might impact on both drug discovery and vaccine design.

Virulence genes

The traditional view of virulence has focused on microbial cell surface structures and secreted toxins that mediate interactions with host cells or tissue damage. However, through genetic analysis it has become apparent that bacterial pathogenesis is a much more complex process with many examples of specific, interdependent interactions between hosts and pathogens[4]. Bacteria can sense and respond to intracellular and extracellular habitats in their hosts, modulate cell surface structures that are the target of immune responses, and scavenge nutrients from their environment to allow replication. The broad view of virulence encompasses all those attributes that contribute to bacterial survival and replication *in vivo*, as well as determinants of tissue tropism and damage.

Microbial genomics

The rapid completion of the *Haemophilus influenzae* genome sequencing project in 1995[5] demonstrated that bacterial sequencing projects can be undertaken by single academic or commercial institutes through improvements in DNA sequencing technology (including sequence assembly), and by avoiding the labour-intensive construction of high-density genetic maps. Since then, there has been an exponential increase in completed genomes with the possibility that the nucleotide sequence for most major bacterial pathogens will be known by the year 2000. The relevance of sequencing projects for drug and vaccine discovery are obvious. The complete sequence includes information on every virulence gene and all potential vaccine candidates, and the sequence databases will become indispensable for research in microbiology. The next challenge is to understand the function of the genes. Unfortunately, *Bacillus subtilis* apart, there do not appear to be any collaborative projects aimed at systematically analysing the data generated from bacterial sequencing projects.

Once the nucleotide sequencing is completed and the sequence assembled, potential open reading frames (orfs) are identified and used to perform database searches for genes of known function in other organisms. In this way, functions are ascribed to orfs and the sequence is annotated. Annotation can also be based on comparing the predicted folding of putative proteins[6], though this is a relatively insensitive method given current knowledge about protein structure. For highly conserved gene products involved in fundamental cellular processes (such as DNA replication and protein synthesis), the assignment of

Table 2 Proportion of predicted proteins of unknown function in bacterial pathogens for which the complete nucleotide sequence is known

Species	Genome size (Mbp)	No. of proteins	Genes without predicted function (%)
Borrelia burgdorferi	0.91	853	41
Escherichia coli	4.64	4288	38
Haemophilus influenzae	1.83	1703	40
Helicobacter pylori	1.7	1590	31
Mycobacterium tuberculosis	4.41	3942	60
Staphylococcus aureus	2.8	2400	>50
Streptococcus pneumoniae	2.5	2300	>50

function from database searches is relatively secure. However, for the majority of orfs, information from database searches is far less certain as annotation frequently ascribes only potential biochemical functions to the gene products. To establish unambiguously the role of gene products in living bacteria requires **genetic** rather than genomic analysis. Overinterpretation of sequence data is more of a problem in those organisms that contain many genes whose products perform biochemically similar reactions[7]. For example, 79 potential ABC transporters have been identified in *Escherichia coli*, but the function of most of these have not been defined. Indeed, 38% of all genes in *E. coli* belong to such paralogous gene families[8]. The frequency of paralogues in bacterial genomes varies from species to species, and is lower in *H. influenzae* and *Helicobacter pylori* than *E. coli*. Given the lack of specific information from database searches, attempts to generate computer models of living cells, based largely on the results of nucleotide sequence matches, are over ambitious and premature.

The bacterial sequencing projects have underscored our limited understanding of the biology of bacteria. In *E. coli*, which has been studied for over 30 years and for which there is more accumulated biochemical and structural information than for any other free-living organism, the function of over 30% of genes is not known. A significant proportion of orfs in every sequenced microbial genome is of unknown function (Table 2). The function of these genes will have to be determined by experimentation, and the importance of mutational analysis cannot be overstated. Genetic research is moving from phenotype-based gene finding to gene-based function finding. The relevant phenotypes relating to drug discovery are bacterial survival and proliferation in the host, and/or the ability to induce tissue damage. In vaccine development, genes must encode proteins involved in the synthesis of immunogenic structures (both of protein and non-protein antigens) or be required for virulence (for live attenuated vaccines).

There is no shortage of potential applications of genome sequence data for the development of drugs and vaccines. These include exploiting the sequence data directly by DNA immunisation[9]. With the sequence data, orfs can be amplified by PCR and ligated directly into the DNA immunisation vectors. DNA vaccination should prove most valuable for bacteria that are predominantly intracellular in the host and that elicit cellular immune responses (*e.g. Chlamydia*, *Mycobacterium tuberculosis* and *Salmonella* spp.). The validity of this approach was demonstrated by expression library immunisation[10] developed with *Mycoplasma pneumoniae*. This method involves cloning random fragments of bacterial DNA in a vector downstream of a promoter active in eukaryotic cells, vaccinating animals with libraries of recombinant plasmids, then challenging animals with *Mycoplasma*. Libraries that confer protection can then be further analysed to identify the clones responsible for protection. Alternative delivery systems for DNA vaccines include live attenuated *Salmonella typhimurium* strains that have been shown to be highly effective carriers of DNA vaccines against *Listeria monocytogenes*[11], and mucosal immunisation with liposomes containing plasmid vectors[12].

Analysis of complete nucleotide sequence for potentially membrane-spanning proteins can identify surface expressed antigens for incorporation in vaccines. Several computer programmes such as TopPred[13] and Moment[14] have been devised to identify membrane-spanning domains; in *Borrelia burgdorferi*, the causative agent of Lyme disease, a total of 526 proteins have one or more putative membrane-spanning domain with 183 having more than one domain[15]. The *Borrelia* sequence also illustrates the potential antigenic diversity of this organism. *Borrelia* can express an extensive variety of lipoproteins (VMPs) on its cell surface, and the genetic mechanism underlying this antigenic switching is well understood[16]; structural genes are translocated from 'silent' loci to an 'expression' site on linear plasmids. The complete sequence has demonstrated the full antigenic repertoire of VMPs, confirming that VMP-based vaccines would be difficult to develop. Many micro-organisms have developed ways to generate antigenic variation which subverts host immune responses[17,18]. Epitopes that are usually non-immunogenic would be expected to be conserved, and vaccination strategies could be employed (*e.g.* improved presentation or mode of delivery) that lead to enhanced responses against these epitopes.

Although the genome sequences do not offer such a direct route to non-protein based vaccines, they can provide invaluable information for the analysis of non-protein vaccine candidates as the basis for the rational design and production of candidate immunogens. An excellent example is the study of *H. influenzae* lipopolysaccharide (LPS). LPS is important both for the interaction of this organism with epithelial cells in the nasopharynx, and as a mediator of septic shock that is sometimes seen during systemic infection[19]. The *Haemophilus* genome was searched for homologues of gene products involved in the biosynthesis of LPS in other organisms[20]. The *H. influenzae* genes responsible for the biosynthesis of terminal LPS structures were identified and mutations introduced into the orfs. The specific mutants were used to analyse the function of each orf by examining their reactivity with a selection of monoclonal antibodies, their LPS structure by mass spectroscopy, and their behaviour in an animal model of infection. By this approach it should be possible to determine firstly precise structure:function relationships for this important molecule, and secondly LPS structures that are safe and potentially immunogenic.

Another valuable aspect of complete sequence information is the identification of chromosomal substrates for the transfer and acquisition of genes involved in either virulence or resistance to antibiotics. The entire repertoire of transposons, phages, and retrons that can harbour cassettes encoding resistance to antimicrobials will be shown by the nucleotide sequence[21]. Important determinants of resistance such as the *van*A operon, responsible for high-level vancomycin resistance in *E. faecium* and *Staph. aureus*, reside on the transposon Tn*1546*[22]. Furthermore, genome analysis will identify whether specific virulence determinants lie on transferable genetic elements. Examples of virulence genes located on resident bacteriophages include those encoding cholera and the diphtheria toxins[23,24]. Additionally, a number of pathogens possess large regions of DNA, often in excess of 10 kb, containing clusters of virulence related genes. These 'pathogenicity islands' were first identified by Hacker in two clinical isolates of uropathogenic *E. coli*[25,26]. The occurrence of pathogenicity islands is of considerable interest, not only because they contain many virulence-related genes, but also because their G+C content usually differs from the remainder of the resident chromosome and the finding of insertion sequences at their termini both suggest they were acquired by horizontal transfer. Not only do these islands provide a means of rapidly identifying virulence gene clusters but also can help understand the intra- and interspecies transfer of these traits. Comparison of the complete nucleotide sequence of closely related bacteria that behave differently in hosts should reveal genes responsible for their difference in behaviour. For instance, *E. coli* O157 (an important cause of bloody diarrhoea and haemolytic uraemic syndrome) has 1 Mb more genetic information than the sequenced *E. coli* K12 strain[27]. One pathogenicity island has already been found in *E. coli* O157 that is absent from *E. coli* K12[28], and the remainder of the O157-specific sequence is located in about 300 regions scattered throughout the genome, some of which will contribute to the pathogenicity of this serotype.

Screens for genes

Virulence genes can be identified using assays that replicate one or more stages of infection. Much work has focused on the interaction of pathogens with monolayers of cells in tissue culture; these screens have facilitated the analysis of bacterial determinants of host cell invasion or intracellular survival. The advantages of working with such *in vitro* systems are that they are easily controlled and amenable to detailed biochemical analysis.

However, *in vitro* screens will only identify a subset of the virulence genes required for pathogenesis because they do not reflect the diverse environments that bacteria encounter in a host. More complex assays are required for a fuller appreciation of bacterial virulence. Recently, two genetic schemes have been developed in which pools of bacterial strains can be screened in animal models of infection. *In vivo* expression technology (IVET) was designed to identify promoters of genes that are specifically induced in host tissues[29], and signature-tagged mutagenesis (STM) allows a large number of insertional mutant strains to be screened simultaneously for loss of virulence in a single animal[30].

IVET is a promoter trap in which random fragments of bacterial DNA are ligated upstream of a promoterless gene whose activity is easily monitored[29,31] (Fig. 1A). A pool of bacteria harbouring plasmids with promoter fusions is inoculated into an animal, and promoters that are specifically active in the host identified. Once these promoters have been isolated, specific mutations are introduced into the corresponding genes. Evaluation of the virulence of the mutant strains is used to confirm whether the genes are required for pathogenesis. IVET of *S. typhimurium,* a Gram-negative bacterium that causes a systemic illness in mice similar to human typhoid fever, resulted in the isolation of three genes that were subsequently shown to be important virulence determinants by mutational and LD_{50} analysis.

Application of IVET to *Pseudomonas aeruginosa* led to the identification of genes induced after intraperitoneal inoculation of neutropenic mice. A total of 22 *in vivo* induced loci were partially sequenced[32]. These encode products involved in iron acquisition, amino acid biosynthesis, signal transduction, gene regulation and several with unknown function. Inactivation of a gene with significant similarity to members of the bacterial ferric uptake regulator family resulted in a 100-fold reduction in LD_{50}, demonstrating the importance of this locus in *Pseudomonas* virulence. IVET was also used to isolate virulence genes in *Staph. aureus* including *agr*, a locus that controls the expression of several virulence genes[33]. Mutagenesis of 11 *in vivo* induced loci produced seven strains significantly attenuated in virulence. These studies show that IVET is a useful approach to identify subsets of genes that are involved in bacterial pathogenesis (Fig. 2).

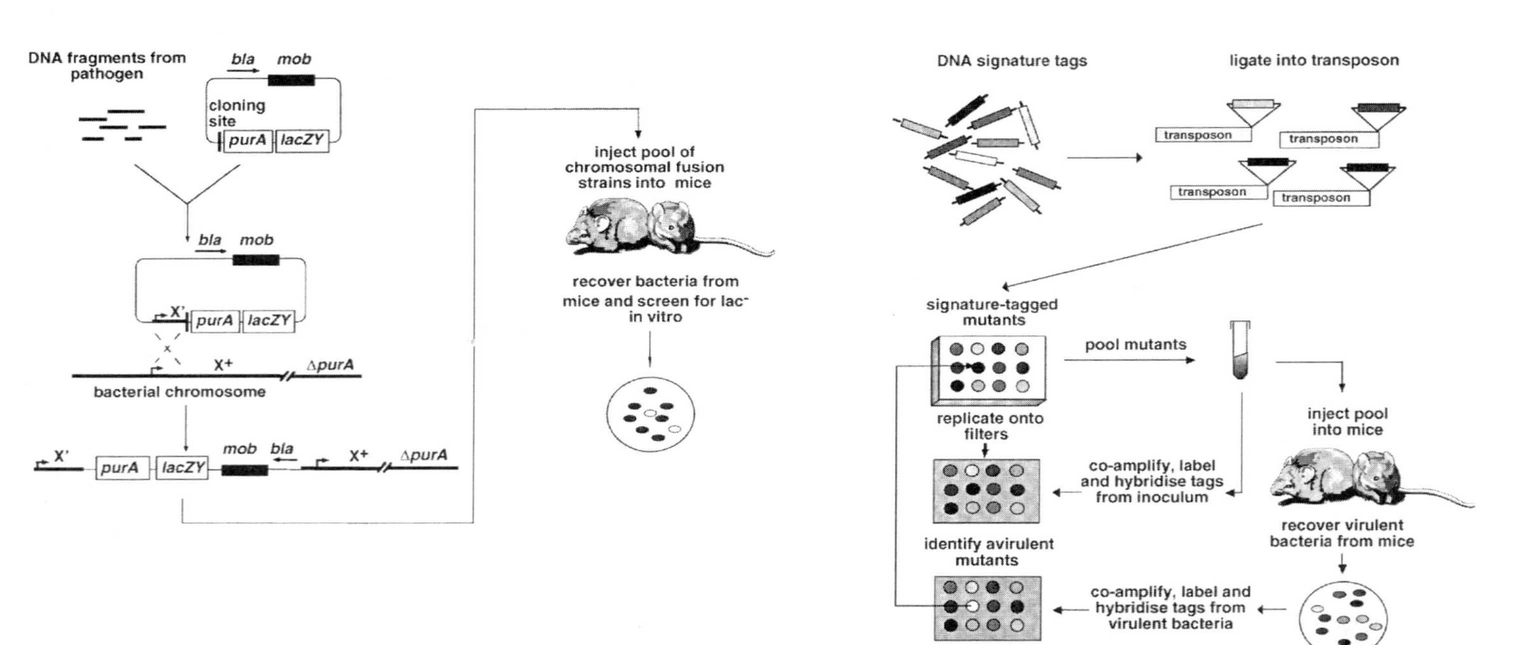

Fig. 1 Principles of IVET and STM. In STM, an insertional mutagen (transposon) is modified by incorporation of DNA signature tags. A collection of different insertional mutants of a bacterial pathogen, each carrying a different tag, is assembled in a microtitre dish. The mutants are pooled and used as the inoculum for an appropriate animal model of infection. After a period of time in which virulent bacteria have multiplied within the host, bacteria are recovered from the host. Signature tags representing strains in the inoculum and strains recovered from the animal are separately amplified using primers that anneal to invariant sequences flanking the signature tags, then labelled and used to probe nylon membranes carrying DNA from the mutants in the microtitre dish. An avirulent mutant is identified by the failure to yield a signal on the membrane hybridized with the tags recovered from the animal. In the original version of IVET, small chromosomal DNA fragments from the bacterial pathogen are cloned upstream of a promoterless operon consisting of *pur*A and *lac*ZY. Introduction of these fusion plasmids into a strain of the pathogen carrying a *pur*A mutation results in their chromosomal integration by homologous recombination. A pool of chromosomal fusion strains is injected into an animal host and, because purine auxotrophy greatly attenuates growth *in vivo*, only those strains in which the *pur*A gene is driven by a bacterial promoter that is active *in vivo* will be pathogenic. The presence of the *lac*ZY gene in the operon allows the transcriptional activity of the promoter to be monitored. Strains that are virulent in the host but Lac- *in vitro* represent *in vivo*-induced (ivi) genes, which can be cloned for further analysis.

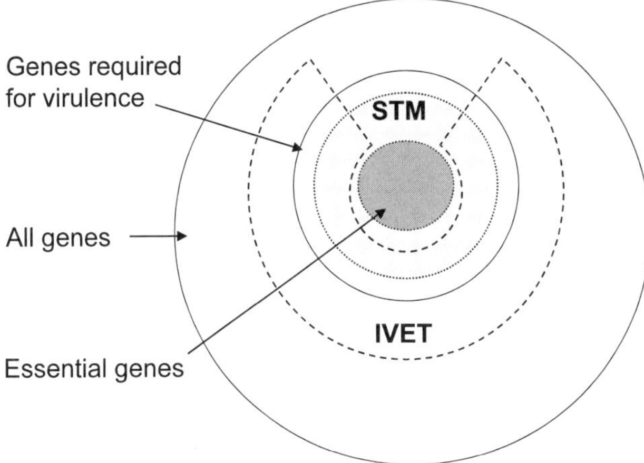

Fig. 2 Categories of bacterial genes identified by STM and IVET. The outer circle encloses all the genes in a bacterial pathogen, and the inner solid circle encloses all genes required for virulence that are potential targets for antimicrobials. STM (pale grey area) identifies a subset of the genes required for virulence; mutations in essential genes (dark grey area) will not be viable, and some mutations, although in genes involved in pathogenesis, will not be attenuated in mixed infections. IVET (dashed outline) identifies genes that are transcribed in the host but not under laboratory conditions; essential genes are transcribed both on laboratory media and in the host.

Signature-tagged mutagenesis (STM) is a technique that allows the fate of a large number of different mutant strains to be monitored simultaneously in a living host (Fig. 1B). In STM, every mutation that is generated carries a different sequence tag that allows mutants to be differentiated from each other. The tags are short segments of DNA containing a 40-bp variable central region flanked by invariant 'arms' that facilitate the co-amplification and labelling of the central portions by PCR. Tagged mutant strains are combined and inoculated into an animal. After infection is established, bacteria are isolated from the animal; mutants that are attenuated will not be recovered from the animals. Comparison of tags that are present in the inoculum but absent from the recovered bacteria identifies attenuated mutants. STM is, therefore, a negative selection technique based on the fact that the majority of random insertional mutations cause a loss-of-function phenotype.

STM of *S. typhimurium* identified both known virulence genes and a novel type III secretion system essential for the systemic growth of this pathogenic microbe. Type III secretion systems consist of a large number of proteins involved in the secretion and translocation of virulence determinants into eukaryotic cells in a contact-dependent manner[34].

Components of type III secretion systems are conserved in *Yersinia*, *Shigella*, *Salmonella*, *Pseudomonas*, and *E. coli,* and are often located on pathogenicity islands. The type III systems are of special interest for drug and vaccine development as they constitute a potential pathogen-specific target present in a range of disease-causing bacteria. Pathogen-specific targets may be preferable to targets also found in commensal species as selection against a sub-set of the microbial flora should make resistance less likely. Furthermore, type III systems could be adapted to deliver heterologous antigens for vaccination or inactivated to generate avirulent bacterial strains as live, attenuated vaccines.

The broader applicability of STM was assessed using the Gram-positive pathogen *Staph. aureus*[35]. *Staph. aureus* causes a wide range of diseases in humans resulting from either localised or systemic infections, leading to abscesses, endocarditis, pneumonia and other life-threatening diseases. For methicillin resistant *Staph. aureus* that is prevalent in both Europe and the US, the only available forms of treatment are the glycopeptides. Reports of the emergence of vancomycin (the principal glycopeptide in clinical use) resistant strains are of great concern[1].

STM was used to identify 50 genes of *Staph. aureus* required for virulence in a murine model of bacteraemia. Analysis of the transposon insertion points in the attenuated mutants revealed that approximately half had insertions in genes of unknown function. Most of the remaining insertion points were in genes involved in nutrient biosynthesis and cell surface metabolism, including *fem*A and *fem*B; a further mutation occurred in a previously unknown gene that shares significant similarity to *fem*B but whose function in cell wall structure is unknown. Both *fem*A and *fem*B are involved in the formation of cell wall peptidoglycan pentaglycine cross-bridges[36]. It is not surprising that such mutants are affected in pathogenesis, because a null mutation in *fem*AB results in a strain that has exclusively monoglycine cross-bridges, a slower growth rate, and aberrant cross-walls[36]. Furthermore, many surface proteins are anchored via the pentaglycine cross-bridges[37], and could be affected in *fem*A or *fem*B mutants.

Both IVET and STM have their limitations as genetic screens. In the IVET screens performed to date, only promoters that are active in the host and not on laboratory media have been analysed. However, this is an arbitrary exclusion criterion. There is no *a priori* reason why genes transcribed both during infection and on laboratory media are not of interest in bacterial virulence; indeed, the adhesins expressed by *Shigella* and *Yersinia,* and the type III secretion systems expressed by *Salmonella* fall into this category. Furthermore, once a promoter with the desired pattern of activity has been isolated, it is still necessary to perform mutagenesis and animal studies to evaluate whether the gene is involved in virulence. STM combines the power of mutational analysis with the ability to screen many mutant strains in a single assay. However, two categories of virulence genes will not be isolated by STM. Firstly, it will not be possible to obtain complete loss-of-function mutations in essential genes whose products might be useful drug targets. Secondly, some mutant strains, for instance those with insertions in genes encoding secreted bacterial toxins, will be attenuated when tested individually but not when inoculated in a pool of toxin producing strains. STM primarily identifies mutant strains that are viable on laboratory media but defective for growth in a host. IVET and STM are therefore complementary technologies that offer the prospect of rapid identification of bacterial virulence genes (Fig. 2).

From virulence genes to drug and vaccine discovery

Mutation of virulence genes frequently produces non-pathogenic strains that are, nevertheless, viable when grown on laboratory media, making them candidates as live, attenuated vaccines. The best examples of live attenuated bacterial vaccines are the two *Salmonella typhi* vaccines that are both immunogenic in human volunteers after administration of a single dose[38,39]. One vaccine strain has a deletion in the *pho*P/*pho*Q sensor kinase system that modulates the expression of a large number of virulence genes[38]. The other vaccine strain carries deletions in genes encoding part of the purine biosynthetic pathway[39]. Therefore, an understanding of the nutritional requirements of bacteria during infection as well as the microbial regulatory networks of virulence genes have proved valuable in vaccine development. Not only can attenuated strains act as vaccines in their own right, they can also be used to deliver heterologous antigens. An attractive goal is to construct live attenuated vaccines carrying antigens from many pathogens that can be administered as a single dose mucosally[40].

The ideal drug target should fulfil the following criteria. The target should: (i) be conserved across a range of pathogenic bacteria; (ii) not be present in eukaryotic cells; (iii) have an activity that can be assayed by *in vitro* tests; and (iv) show an inhibition profile that leads to microbial death or loss of pathogenicity.

Much interest continues to focus on the metabolic pathways that bacteria require during growth in the host. A good example of this is the aromatic amino acid biosynthetic pathway that is present in microbes but not in eukaryotic cells. A number of antimicrobial compounds (*e.g.* sulphonamides, trimethoprim, dapsone) are active against enzymes involved in terminal steps in this pathway. Many useful antimicrobials, such as the β-lactams and the glycopeptides, act on components unique to the bacterial cell membrane, and are selectively active against microbes. There are additional cell surface structures and enzymes that catalyse the synthesis of these structures that could be targets for effective antimicrobials. The significantly greater β-lactam susceptibility of *fem*A and *fem*B mutants has suggested that the FemA and FemB proteins are potential targets for drugs that could restore sensitivity to β-lactam antibiotics, though such drugs may prove to be useful antibiotics in their own right. There is also interest in developing agents that act against bacterial LPS.

Most of the work on drug development has relied upon the ability to assay the activity of potential drug targets and then to screen chemical libraries for compounds that inhibit the activity. This approach has proved extremely valuable in the past, and is the basis for all the classes of antimicrobials in current use. New compounds have been derived by chemical modification of the structure of these classes and re-testing. However, in recent years, repeated screening of compound libraries has led to the re-isolation of the same classes of inhibitors[41]. A new strategy is being developed that relies upon determining the detailed structure of the intended drug target to design specific inhibitors. Ligand binding approaches have not as yet resulted in the licensing of any new drugs, but there is hope in the pharmaceutical industry that this approach will yield dividends in terms of new drugs. A good example of the application of ligand binding is the recent solving of the structure of AroM bound to an inhibitor[42]. AroM is a pentavalent enzyme in aromatic amino acid biosynthesis that is not present in eukaryotic cells. Detailed analysis of the structure of this molecule has led to the synthesis of inhibitors that can block its activity. The demonstration of inhibition *in vitro* is a long way from having an effective antibiotic; compounds must reach the site of infection in the host, enter the bacterial cell, and pass all the stringent safety testing laid down for new compounds. However, the approach marks a way forward in generating novel compounds for evaluation.

Conclusion

The race between drug development and bacterial evolution is currently being led by the microbes. However, the rapid increase in our understanding of molecular mechanisms of bacterial pathogenesis, made possible by the genome sequencing projects and new genetic methods, should provide an abundance of information for drug and vaccine discovery. The next challenge will be to convert this information into practical applications to reduce the burden of bacterial diseases.

References

1 Hiramatsu K, Aritaka N, Hanaki H *et al.* Dissemination in Japanese hospitals of strains of *Staphylococcus aureus* heterogeneously resistant to vancomycin. *Lancet* 1997; **350:** 1670–3

2 Michel M, Gutmann L. Methicillin-resistant *Staphylococcus aureus* and vancomycin-resistant enterococci: therapeutic realities and possibilities. *Lancet* 1997; **349:** 1901–6

3. World Health Organization. *The World Health Report 1996. Fighting Disease Fostering Development.* Geneva: World Health Organization, 1996

4 Finlay BB, Cossart P. Exploitation of mammalian host cell functions by bacterial pathogens. *Science* 1997; **276:** 718–25

5 Fleischmann RD, Adams MD, White O *et al.* Whole-genome random sequencing and assembly of *Haemophilus influenzae* Rd. *Science* 1995; **269:** 496–512

6 Fischer D, Eisenberg D. Assigning folds to the proteins encoded by the genome of *Mycoplasma genitalium*. *Proc Natl Acad Sci USA* 1997; **94:** 11929–34

7 Tatusov RL, Koonin EV, Lipman DJ. A genomic perspective on protein families. *Science* 1997; **278:** 631–7

8 Linton KJ, Higgins CF. The *Escherichia coli* ATP-binding cassette (ABC) proteins. *Mol Microbiol* 1998; **28:** 5–13

9 Ulmer JB, Sadoff JC, Liu MA. DNA vaccines. *Curr Opin Immunol* 1996; **8:** 531–6

10 Barry MA, Lai WC, Johnston SA. Protection against mycoplasma infection using expression-library immunization. *Nature* 1995; **377:** 632–5

11 Darji A, Guzman CA, Gerstel B *et al.* Oral somatic transgene vaccination using attenuated *S. typhimurium*. *Cell* 1997; **91:** 765–75

12 Klavinskis LS, Gao L, Barnfield C, Lehner T, Parker S. Mucosal immunization with DNA-liposome complexes. *Vaccine* 1997; **15:** 818–20

13 Claros MG, von Heijne G. TopPred II: an improved software for membrane protein structure predictions. *Comput Appl Biosci* 1994; **10:** 685–6

14 Eisenberg D, Schwarz E, Komaromy M, Wall R. Analysis of membrane and surface protein sequences with the hydrophobic moment plot. *J Mol Biol* 1984; **179:** 125–42

15 Fraser CM, Casjens S, Huang WM *et al.* Genomic sequence of a Lyme disease spirochaete, *Borrelia burgdorferi*. *Nature* 1997; **390:** 580–6

16 Zhang JR, Hardham JM, Barbour AG, Norris SJ. Antigenic variation in Lyme disease borreliae by promiscuous recombination of VMP-like sequence cassettes. *Cell* 1997; **89:** 275–85

17 Robertson BD, Meyer TF. Genetic variation in pathogenic bacteria. *Trends Genet* 1992; **8:** 422–7

18 Moxon ER, Rainey PB, Nowak MA, Lenski RE. Adaptive evolution of highly mutable loci in pathogenic bacteria. *Curr Biol* 1994; **4:** 24–33

19 Henderson B, Poole S, Wilson M. Bacterial modulins: a novel class of virulence factors which cause host tissue pathology by inducing cytokine synthesis. *Microbiol Rev* 1996; **60:** 316–41

20 Hood DW, Deadman ME, Allen T *et al.* Use of the complete genome sequence information of *Haemophilus influenzae* strain Rd to investigate lipopolysaccharide biosynthesis. *Mol Microbiol* 1997; **22:** 951–66

21 Mazel D, Dychinco B, Webb VA, Davies J. A distinctive class of integron in the *Vibrio cholerae* genome. *Science* 1998; **280:** 605–8

22 Woodford N, Adebiyi AM, Palepou MF, Cookson BD. Diversity of VanA glycopeptide resistance elements in enterococci from humans and nonhuman sources. *Antimicrob Agents Chemother* 1998; **42:** 502–8

23 Matsuda M, Barksdale L. Phage-directed synthesis of diphtherial toxin in non-toxinogenic *Corynebacterium diphtheriae*. *Nature* 1966; **210:** 911–3

24 Waldor MK, Mekalanos JJ. Lysogenic conversion by a filamentous phage encoding cholera toxin. *Science* 1996; **272:** 1910–4

25 Blum G, Ott M, Lischewski A *et al.* Excision of large DNA regions termed pathogenicity islands from tRNA- specific loci in the chromosome of an *Escherichia coli* wild-type pathogen. *Infect Immun* 1994; **62:** 606–14

26 Hacker J, Blum-Oehler G, Muhldorfer I, Tschape H. Pathogenicity islands of virulent bacteria: structure, function and impact on microbial evolution. *Mol Microbiol* 1997; **23:** 1089–97

27 Blattner FR, Plunkett 3rd G, Bloch CA *et al.* The complete genome sequence of *Escherichia coli* K-12. *Science* 1997; **277:** 1453–74

28 McDaniel TK, Jarvis KG, Donnenberg MS, Kaper JB. A genetic locus of enterocyte effacement conserved among diverse enterobacterial pathogens. *Proc Natl Acad Sci USA* 1995; **92:** 1664–8

29 Mahan MJ, Slauch JM, Mekalanos JJ. Selection of bacterial virulence genes that are specifically induced in host tissues. *Science* 1993; **259:** 686–8

30 Hensel M, Shea JE, Gleeson C, Jones MD, Dalton E, Holden DW. Simultaneous identification of bacterial virulence genes by negative selection. *Science* 1995; **269:** 400–3

31 Camilli A, Mekalanos JJ. Use of recombinase gene fusions to identify *Vibrio cholerae* genes induced during infection. *Mol Microbiol* 1995; **18:** 671–83

32 Wang J, Mushegian A, Lory S, Jin S. Large-scale isolation of candidate virulence genes of *Pseudomonas aeruginosa* by *in vivo* selection. *Proc Natl Acad Sci USA* 199; **93:** 10434–9

33 Peng HL, Novick RP, Kreiswirth B, Kornblum J, Schlievert P. Cloning, characterization, and sequencing of an accessory gene regulator (*agr*) in *Staphylococcus aureus. J Bacteriol* 1988; **170:** 4365–72

34 Cotter PA, Miller JF. Triggering bacterial virulence. *Science* 1996; **273:** 1183–4

35 Mei JM, Nourbakhsh F, Ford CW, Holden DW. Identification of *Staphylococcus aureus* virulence genes in a murine model of bacteraemia using signature-tagged mutagenesis. *Mol Microbiol* 1997; **26:** 399–407

36 Stranden AM, Ehlert K, Labischinski H, Berger-Bachi B. Cell wall monoglycine cross-bridges and methicillin hypersusceptibility in a *fem*AB null mutant of methicillin-resistant *Staphylococcus aureus. J Bacteriol* 1997; **179:** 9–16

37 Schneewind O, Fowler A, Faull KF. Structure of the cell wall anchor of surface proteins in *Staphylococcus aureus. Science* 1995; **268:** 103–6

38 Hohmann EL, Oletta CA, Killeen KP, Miller SI. *pho*P/*pho*Q-deleted *Salmonella typhi* (Ty800) is a safe and immunogenic single-dose typhoid fever vaccine in volunteers. *J Infect Dis* 1996; **173:** 1408–14

39 Tacket CO, Sztein MB, Losonsky GA *et al.* Safety of live oral *Salmonella typhi* vaccine strains with deletions in *htr*A and *aro*C *aro*D and immune response in humans. *Infect Immun* 1997; **65:** 452–6

40 Levine MM, Dougan G. Optimism over vaccines administered via mucosal surfaces. *Lancet* 1998; **351:** 1375–6

41 Chopra I, Hodgson J, Metcalf B, Poste G. New approaches to the control of infections caused by antibiotic-resistant bacteria. An industry perspective. *JAMA* 1996; **275:** 401–3

42 Carpenter EP, Hawkins AR, Frost JW, Brown KA. Structure of dehydroquinonate synthase reveals an active site capable of multiple catalysis. Nature 1998; 394: 290–302

Genetics and genomics of infectious disease susceptibility

Adrian VS Hill

Wellcome Trust Centre for Human Genetics, Nuffield Department of Clinical Medicine, University of Oxford, Oxford, UK

Human genetic variation is a major determinant of susceptibility to many common infectious diseases. Malaria was the first disease to be studied extensively and many susceptibility and resistance loci have been identified. However, genes for other diseases such as HIV/AIDS and mycobacterial infections are now being identified using a variety of approaches. A large number of genes appear to influence susceptibility to infectious pathogens and defining these can provide insights into pathogenic and protective mechanisms and identify new molecular targets for prophylactic and therapeutic interventions. Immunogenetic associations with infectious diseases have considerable potential to guide immunomodulatory interventions and vaccine design.

Although it has been clear for many years that individuals may differ markedly in their susceptibility to infectious diseases, recent advances in genomics have led to a dramatic increase in the power of techniques available to identify the relevant genes. Some infectious diseases were once regarded as familial before the identification of the causative micro-organism[1] and early twin studies found that there was a substantial host genetic influence on susceptibility to diseases such as tuberculosis and polio. Today, it is clear that human genetic variation exerts a major influence on the course of disease caused by many infectious micro-organisms. Such host–pathogen gene interactions are of general biological interest as they underlie the maintenance of much genetic diversity and such co-evolutionary interplay is often best studied in human infectious diseases where both the pathogen and the host genome are well characterised.

In this short chapter, I shall first review some of the powerful approaches to analysing the host genetic contribution to infectious disease susceptibility that are now available and in use as a result of recent dramatic progress in characterising the human genome. Secondly, I shall attempt to summarise briefly the state of knowledge on genetic factors influencing susceptibility to some major infectious diseases. This selection is a personal one and not at all comprehensive, but may

Correspondence to: Dr Adrian VS Hill, Wellcome Trust Centre for Human Genetics, Nuffield Department of Clinical Medicine, University of Oxford, Roosevelt Drive, Oxford OX3 7BN, UK

illustrate the great potential of new molecular genetic approaches to characterising the human genetic basis of variable susceptibility to infection. Although outside the scope of this review, in parallel with progress in host genetics important strides are being made in the genomics of many micro-organisms and, together, these complementary genetic approaches hold considerable promise for explaining variability in the outcome of host–pathogen encounters.

Mapping and identifying the relevant genes

The most frequently adopted approach in the human genetics of infectious disease has been the assessment of candidate genes in case-control studies. However, a variety of different approaches to gene mapping and identification may now be employed in studies of susceptibility to complex disease and most of these have been applied to at least one infectious disease. In general, two distinct approaches may be used. Either a genetic linkage study may be undertaken to search for co-segregation of a genetic marker with disease in families, or a genetic association study is used to determine whether an allele is found more or less frequently in individuals with the disease than in unaffected controls. There are advantages and disadvantages to each approach and ideally both should be employed.

For association studies, large sample sizes are generally required, particularly to assess rare alleles or multi-allelic genes. Control populations need to be carefully matched to the cases to avoid false-positive or false-negative associations due to population stratification. The most important variables to match are ethnic group and locality of birth and residence. Matching for age and sex is usually less important as gene frequencies change little with age and autosomal alleles seldom show sex differences in frequency. Parental controls may be used to avoid the possibility of unknown confounding variables differing between case and controls[2]. A large number of candidate genes are now available for study and these have been suggested by a variety of approaches. The geographic distribution of haemoglobin gene variants in the Mediterranean first suggested that they might play a role in malaria resistance[3]. More recently, some loci, such as the Mx influenza resistance gene, have been found to affect susceptibility to infection in different strains of mice leading to assessment of their human homologues[4]. However, a much larger number of candidate genes have now been suggested by studies of the susceptibility of various gene knockout mice to infectious pathogens. Perhaps the largest class of candidate genes includes those implicated by a knowledge of the

Table 1 A selection of susceptibility and resistance genes implicated in infectious diseases.

Gene	Variant	Disease	Effect
ABO	Blood group O	Cholera	Susceptibility
α-Globin	Thalassaemias	Malaria (Pf)	Resistance
β-Globin	Sickle, thalassaemias	Malaria (Pf)	Resistance
Erythrocyte band 3	27 bp deletion	Malaria (Pf, Pv)	Resistance
G6PD	Deficiency variants	Malaria (Pf)	Resistance
HLA-B	HLA-B53	Malaria (Pf)	Resistance
HLA-DR	HLA-DRB1*1302	Malaria (Pf)	Resistance
HLA-DR	HLA-DR2	Tuberculosis	Susceptibility
HLA-DR	HLA-DR2	Leprosy	Susceptibility
HLA-DR	HLA-DRB1*1302	HBV persistence	Resistance
HLA-DR	HLA-DRB1*11	HCV persistence	Resistance
TNF	Promoter 308	Malaria (Pf)	Susceptibility
FUT2	Non-secretors	UTI	Susceptibility
NRAMP1	5′ and 3′ variants	Tuberculosis	Susceptibility
Interferon-γ receptor	Various mutations	Disseminated BCG	Susceptibility
IL-12 receptor	Various mutations	Intracellular bacteria	Susceptibility
CCR5	32 bp deletion	HIV infection/progression	Resistance
CCR2	codon 64	HIV progression	Resistance
Duffy receptor	Promoter variant	Malaria (Pv)	Resistance
PRP	Codon 129	Creutzfeldt-Jakob Disease	Susceptibility

Pf, *Plasmodium falciparum*; Pv *Plasmodium vivax.*

pathophysiology and immunology of the particular disease. For many infectious diseases this list would include genes such as HLA, tumour necrosis factor and mannose-binding lectin variants which have been studied frequently because of their known functions (Table 1).

A newer approach in human infectious disease genetics is to search for genetic linkage to, rather than association with, a disease in family studies. With the completion of a genetic map of the human genome and the identification of thousands of very variable microsatellite markers it has become possible to search the whole genome efficiently for regions linked to disease susceptibility. Thus, identification of a genetic marker convincingly linked to susceptibility indicates that there is a susceptibility gene somewhere in that chromosomal region[5]. Typically, hundreds of families with two affected siblings need to be studied to provide compelling evidence of linkage. Chromosomal regions identified in this way are initially very large, and contain some hundreds of genes distributed over several megabases, so that much further work is usually required to actually identify the causative gene. However, the attraction of this relatively laborious approach is that unknown genes may be mapped and identified without prior information on their function. The main concern about the applicability of this approach to common infectious diseases is that the statistical power of this approach is generally lower than that of case-control studies, in part because fewer multi-case families than random cases can be recruited. Thus, if the genetic contribution to susceptibility is accounted for by a large number of genes with modest or small effects, they may not be picked up on linkage analysis. Nonetheless, this approach does allow a comprehensive screen of the whole genome to be undertaken to search for any major susceptibility genes, and major genes may be those of most biological interest. Similar linkage approaches have been used in mice to map the locations of many susceptibility genes[6] and one of these, termed *Nramp1*, a susceptibility gene for *BCG, Leishmania* and *Salmonella* in mice, was isolated by positional cloning[7]. The human homologue of this gene, *NRAMP1*, has recently been found to affect susceptibility to pulmonary tuberculosis[8] in West Africans and may also be involved in susceptibility to diseases caused by other intracellular pathogens.

Diseases

Mycobacterial diseases

Leprosy and tuberculosis have been frequently studied by human geneticists for several reasons. Both show familial clustering and, being chronic diseases which require a protracted period of treatment, it is relatively easy to recruit large numbers of cases and even multi-case families. Also, there have been twin studies of both diseases that found substantially higher concordance rates in monozygotic than dizygotic twins[9,10], and thus provided an estimate of the magnitude of the host genetic component to variable susceptibility. Early studies of HLA variation in India and Surinam found associations with both tuberculosis and leprosy[11,12]. HLA-DR2 (particularly the HLA-DR15 subtype) was associated with susceptibility to tuberculoid leprosy in India and more recent data show an association of this HLA type with susceptibility to both tuberculoid and lepromatous forms of leprosy as well as to tuberculosis in several Asian populations[13–16]. However, in other continents no clear HLA association has been identified and HLA-DR2 appears not to be associated with susceptibility. Recently, in a study of leprosy in Calcutta variation in the promoter of the tumour necrosis factor gene was associated with susceptibility to lepromatous but not tuberculoid leprosy[17]. A variant of the vitamin D receptor has also been associated with the type of leprosy developed in the same study population (Roy *et al.* manuscript submitted).

In a study in The Gambia, West Africa, allelic variation in both the 3′ untranslated region and in the promoter region of the natural resistance-associated macrophage protein-1 (*NRAMP1*) gene was associated with susceptibility to sputum-positive pulmonary tuberculosis[8]. A genome-wide linkage analysis of tuberculosis in Africans has now been undertaken and this has suggested that there may be genes with relatively major effects on susceptibility located on chromosomes 15 and X (Bellamy *et al.* manuscript submitted). However, much further work will be required to identify these putative susceptibility genes. Analyses of rare individuals with marked susceptibility to atypical mycobacterial disease has recently revealed several interesting molecular defects. Some such children who are homozygous for mutations in the interferon-γ receptor gene are remarkably susceptible to weakly pathogenic mycobacteria, including the BCG vaccine, and have a poor prognosis[18,19]. But whether these children have increased susceptibility to tuberculosis and leprosy is not known. Similarly, 'knockout' mutations in the IL-12 receptor beta-1 gene produce a phenotype of marked susceptibility to atypical mycobacterial disease[20,21] and, in one family, to *Salmonella* infections. Although such mutations that have major effects on mycobacterial susceptibility must be rare, it is possible that milder defects in these cytokine and receptor genes might explain some general variation in susceptibility to tuberculosis.

Malaria

Malaria provides the classic examples of infectious disease resistance genes and more genes have been implicated in differential susceptibility to malaria than to any other infectious disease. In the 1930s, studies of the use of malaria therapy in the management of syphilis suggested some marked interindividual differences in susceptibility to malaria in non-immunes[22]. The subsequent identification of the major effect of sickle haemoglobin heterozygosity on malaria resistance[23] and the emerging picture of the geographical distribution of some haemoglobin variants provided the incentive for genetic investigations in a diversity of populations. Such studies of sickle haemoglobin and G6PD deficiency have provided clear-cut evidence of their protective relevance against *Plasmodium falciparum* malaria[24,25], but it is still uncertain whether haemoglobin C, which is common in parts of West Africa, and haemoglobin E, widely distributed in Southeast Asia, are protective. A few studies have demonstrated protection associated with either heterozygosity for β-thalassaemia or, more recently, various α-thalassaemia genotypes, but the mechanisms of protection remain uncertain[26,27]. Interestingly, it has recently been suggested that some of this protection may have an immune basis and that interactions between susceptibility to *Plasmodium vivax* and *P. falciparum* may be relevant in populations where both are prevalent[28].

Despite the geographical variation in frequencies of many malaria resistance alleles, there have been few useful interpopulation comparisons of malaria susceptibility. However, a study of different ethnic groups in Mali, West Africa has found significant differences in immune responses to *P. falciparum* and in malaria susceptibility between these groups that appear to be genetic in origin[29]. These differences could not be explained by the known malaria resistance alleles.

Both HLA class I and II alleles have been found to influence malaria susceptibility in Africa in large case-control studies[30,31]. In the largest study in The Gambia, HLA-B53 was associated with resistance to both cerebral malaria and severe malarial anaemia, whereas the class II allele, HLA-DRB1*1302, was associated with resistance only to the latter. Subsequent immunological investigations in this population have suggested a possible molecular basis for these HLA associations through the identification of peptide epitopes in the parasite restricted by these HLA types (see below)[32]. In a study in Kenya, a different HLA class II type was associated with protection, and interpopulation heterogeneity in HLA associations appears to be fairly common in infectious diseases. This can have many causes but a prominent one in malaria is likely to be the marked polymorphism of immunodominant malaria antigens. HLA has recently been found to influence the strain of malaria parasite associated with clinical malaria and complex interactions between malaria parasite strains may lead to further variability in HLA associations[33]. Geographical heterogeneity in association may also be found for other genes. For example a polymorphic host receptor involved in parasite sequestration, intercellular adhesion molecule-1 (ICAM-1), may influence susceptibility to cerebral malaria[34]. Homozygotes for an African-specific variant, ICAM-Kilifi, were found significantly more frequently amongst cases of cerebral malaria in Kenya, but in The Gambia no effect of this variant on malaria susceptibility was detected (Bellamy *et al.* 1998 in press).

HIV and AIDS

There has been considerable interest in investigating genetic susceptibility to HIV and AIDS over the last few years. This has been encouraged by studies of cohorts in which a small proportion of individuals remain HIV seronegative despite repeated exposure to HIV from infected sexual partners[35]. Immunological assays have confirmed that some such resistant sex workers have been exposed to the virus. Individuals also vary in their rate of disease progression to AIDS once infected and several genes have now been identified that appear to influence this rate. HLA studies have shown that in several populations HLA-B35 and the HLA-A1-B8-DR3 haplotype are associated with more rapid disease progression[36,37] and HLA-B27 and HLA-B57 with a lower rate of progression[38]. Some combinations of HLA class I and II alleles and variants of the transporter associated with antigen processing (TAP) genes have also been implicated[39] in north American cohorts. A genetic linkage study of haemophiliac brothers also demonstrated an effect of major histocompatibility complex variation on rate of CD4+ T cell decline[40]. Thus, despite the variability of HIV between and within individual infections, HLA type has been found to be a significant if minor determinant of the rate of disease progression.

Recently, there have been numerous genetic analyses of chemokine receptors that are coreceptors with CD4 for viral entry into macrophages and lymphocytes. Homozygotes for a 32 bp deletion in the gene encoding CC chemokine receptor-5 (CCR5), the coreceptor for macrophage-tropic HIV, are very markedly resistant to HIV infection and heterozygotes display lower rates of disease progression[41]. Variants of the flanking gene for the CCR2 chemokine receptor and of the stromal-derived factor (SDF-1) gene encoding the ligand for CXCR4, the coreceptor for lymphocyte-tropic strains, have also been associated with some alteration in rate of disease progression[42,43]. However, several other resistance genes must exist as the known variants of CCR5 account for only a minority of Caucasians and none of the African individuals found to be markedly resistant to HIV infection.

Other infectious diseases

Many genetic associations with other infectious diseases have been reported and will probably turn out to be important in multiple populations. For example, genetic variation in the mannose-binding ligand gene[44] is likely to influence susceptibility to several bacterial pathogens. A codon 129 polymorphism in the host prion protein (PRNP) gene is strongly associated with susceptibility to both iatrogenic Creutzfeldt-Jakob Disease (CJD) in US and French series and to new-variant CJD in UK cases[45,46]. The genetic linkage approach has recently been successfully applied to map a gene affecting susceptibility to the helminth, *Schistosomiasis mansoni*, in Brazilian families[47]. The chromosome 5 region identified encodes numerous cytokine genes such as IL-4 and IL-13 and this gene cluster has also been linked to atopy and asthma, consistent with the speculation that a gene selected to provide resistance to helminthic infections might predispose to asthma or atopy. Finally, the recent finding that the cystic fibrosis transmembrane conductance regulator (CFTR) is the intestinal receptor for *Salmonella typhi*[48] raises the possibility that common cystic fibrosis mutations may have been selected in Caucasians to provide resistance to typhoid.

Applications

We are only beginning to uncover what will probably turn out to be a huge number of genes involved in variable susceptibility to infectious disease. There are many reasons for continuing to define these genetic factors and elucidate their mechanisms.

Risk assessment

One application will be in risk prediction and this genetic information will probably influence behaviour, travel patterns, and the use of prophylactic anti-microbials and immunisations. Genetic profiling to estimate individual susceptibility will have a place and already screening for mannose-binding ligand deficiency alleles has been advocated, but the most useful risk profiling will probably involve typing of numerous genetic loci.

Mechanisms in pathogenesis and protection

Genetic associations have already provided numerous insights into the pathogenesis of infectious disease and the relevance of particular defence mechanisms. Association of polymorphisms in cytokines and chemokines or their receptors has led to attempts to modulate the activity of these mediators in particular diseases. For example, the up-regulatory variant of the polymorphism at position –308 of the TNF promoter[49] was associated with susceptibility to severe malaria[50] and agents that may reduce the activity of this cytokine are under assessment[51,52].

Another application is in the understanding of specific immune defences used in host resistance to infection or disease. For example, the enhanced susceptibility to non-virulent mycobacteria in children with mutations in the interferon-γ receptor has highlighted the importance of this pathway in controlling these mycobacteria. But the finding that these children appear to have little or no alteration in their susceptibility to other common pathogens has also been informative. Studies of mannose-binding ligand deficiency have demonstrated that this molecule plays a key role in resistance to some but not to other infectious agents[53]. A major goal in the field is the identification of genetic loci that influence the predominant type of cellular immune response to infectious pathogens and potential allergens. Substantial progress has been made in mapping genes affecting atopy and allergy[54] and these may also be relevant to infectious disease. Candidates for influencing the TH1–TH2 shift of the cellular immune responses are numerous and include IL-4, its receptor, several other cytokines and the vitamin D receptor.

Vaccine design

Immunogenetic associations with infectious diseases may facilitate vaccine design. For example the association between the HLA class I molecule, HLA-B53, and resistance to severe malaria in African children[30] has been analysed in detail. This association was not secondary to known variation in flanking HLA class II or class III region genes[30,50]. This suggested that the association resulted from the action of HLA class I restricted T cells. A search for such cells in Africans exposed to malaria employed a strategy known as reverse immunogenetics[32,55]. Peptides are eluted from the groove of the disease-associated HLA molecule and sequenced to identify the characteristics of peptides that bind to that HLA type. This sequence information is then used to scan protein sequences from the relevant micro-organism to identify candidate epitopes for T cells restricted by that HLA molecule[56]. Application of this approach in malaria led to the identification of an epitope restricted by HLA-B53 in the *P. falciparum* liver-stage antigen-1[32]. This was the first short epitope for CD8+ cytotoxic T cells to be defined in the malaria parasite and it is very substantially conserved across different strains of this very variable pathogen. These findings provided support for efforts to develop vaccines against malaria that would induce protective CD8+ T cells against liver-stage antigens[57]. This vaccine strategy has come up against the problem of the weak immunogenicity of most conventional vaccines for CD8+ T cell induction. However, recently a new prime-boost immunisation strategy may have overcome this difficulty and clinical trials of a new generation of CD8+ T cell-inducing vaccines against malaria are in progress[58]. Animal models of malaria infection have also provided support for this vaccination approach against liver-stage antigens. However, because the pathogen of humans, *P. falciparum*, does not infect rodents or most primates, the immunogenetic data on human malaria have been of particular value.

Several HLA class II associations with infectious diseases have also been defined[59]. These likely reflect an important role for particular CD4+ T cells in immune defence and a similar reverse immunogenetic approach has been taken for the HLA-DR13 association with resistance to chronic hepatitis B virus infection[60]. However, there is additional complexity in analysing these class II associations because of the considerable functional diversity of CD4+ T cells. Definition of associations with other immunoregulatory genes as well as HLA class II type may be valuable in defining the particular type of CD4+ T cell that is of protective relevance.

New pharmacological targets

Perhaps the most important application in future will be the identification of new molecules and pathways which may become targets for pharmacological intervention. A recent illustration of the rapid application of a finding in host genetics comes from the HIV field. Demonstration of the very substantial resistance to HIV infection of homozygotes for a deletion in the CCR5 gene has underpinned ongoing attempts to develop pharmacological blockers of this viral coreceptor. Other potential coreceptors for HIV, such as CCR2, are now the subject of detailed genetic studies[61] as are potential receptors and coreceptors for many other infectious pathogens. Newer techniques of genome-wide analysis using association analysis[62] as well as the established linkage approaches offer the prospect of many new target molecules in these highly polygenic diseases. The power of these new tools has generated considerable excitement in this field and it is likely that what has been discovered so far is only the tip of an iceberg of informative genetic information on disease aetiology.

Acknowledgements

I thank my numerous colleagues and collaborators in the UK and other countries whose work, results and ideas have contributed to this review. The author is a Wellcome Trust Principal Research Fellow.

References

1 Harboe M. Gerhard Henrik Armauer Hansen – still of current interest. *Tidsskr Nor Laegeforen* 1992; **112**: 3795–8
2 Flanders WD, Khoury MJ. Analysis of case-parental control studies: method for the study of associations between disease and genetic markers. *Am J Epidemiol* 1996; **144**: 696–703
3 Haldane JBS. Disease and evolution. *La Ricercha Scientifica* 1949; **19 Suppl**: 68–76
4 Staeheli P, Grob R, Meier E, Sutcliffe JG, Haller O. Influenza virus-susceptible mice carry Mx genes with a large deletion or a nonsense mutation. *Mol Cell Biol* 1988; **8**: 4518–23
5 Weeks DE, Lathrop GM. Polygenic disease: methods for mapping complex disease traits. *Trends Genet* 1995; **11**: 513–9
6 McLeod R, Buschman E, Arbuckle LD, Skamene E. Immunogenetics in the analysis of resistance to intracellular pathogens. *Curr Opin Immunol* 1995; **7**: 539–52
7 Vidal SM, Malo D, Vogan K, Skamene E, Gros P. Natural resistance to infection with intracellular parasites: isolation of a candidate for BCG. *Cell* 1993; **73**: 469–85

8 Bellamy R, Ruwende C, Corrah T, McAdam KP, Whittle HC, Hill AVS. Variations in the NRAMP1 gene and susceptibility to tuberculosis in West Africans. *N Engl J Med* 1998; **338**: 640–4

9 Chakravarti MR, Vogel F. A twin study on leprosy. *Top Hum Genet* 1973; **1**: 1–123

10 Comstock GW. Tuberculosis in twins: a re-analysis of the Prophit survey. *Am Rev Respir Dis* 1978; **117**: 621–4

11 de Vries RR, Fat RF, Nijenhuis LE, van Rood JJ. HLA-linked genetic control of host response to *Mycobacterium leprae*. *Lancet* 1976; ii: 1328–30

12 Singh SP, Mehra NK, Dingley HB, Pande JN, Vaidya MC. Human leukocyte antigen (HLA)-linked control of susceptibility to pulmonary tuberculosis and association with HLA-DR types. *J Infect Dis* 1983; **148**: 676–81

13 Rani R, Fernandez-Vina MA, Zaheer SA, Beena KR, Stastny P. Study of HLA class II alleles by PCR oligotyping in leprosy patients from north India. *Tissue Antigens* 1993; **42**: 133–7

14 Brahmajothi V, Pitchappan RM, Kakkanaiah VN *et al.*. Association of pulmonary tuberculosis and HLA in south India. *Tubercle* 1991; **72**: 123–32

15 Todd JR, West BC, McDonald JC. Human leukocyte antigen and leprosy: study in northern Louisiana and review. *Rev Infect Dis* 1990; **12**: 63–74

16 Bothamley GH, Beck JS, Schreuder GM, de Vries RR, Kardjito T, Ivanyi J. Association of tuberculosis and *M. tuberculosis*-specific antibody levels with HLA. *J Infect Dis* 1989; **159**: 549–55

17 Roy S, McGuire W, Mascie-Taylor CG *et al.* Tumor necrosis factor promoter polymorphism and susceptibility to lepromatous leprosy. *J Infect Dis* 1997; **176**: 530–2

18 Newport MJ, Huxley CM, Huston S *et al.* A mutation in the interferon-gamma-receptor gene and susceptibility to mycobacterial infection. *N Engl J Med* 1996; **335**: 1941–9

19 Jouanguy E, Altare F, Lamhamedi S *et al.* Interferon-gamma-receptor deficiency in an infant with fatal bacille Calmette-Guèrin infection. *N Engl J Med* 1996; **335**: 1956–61

20 Altare F, Durandy A, Lammas D *et al.* Impairment of mycobacterial immunity in human interleukin-12 receptor deficiency. *Science* 1998; **280**: 1432–5

21 Jong R, Altare F, Haagen IA *et al.* Severe mycobacterial and salmonella infections in interleukin-12 receptor-deficient patients. *Science* 1998; **280**: 1435–8

22 James SP, Nicol WD, Shute PG. A study of induced malignant tertian malaria. *Proc R Soc Med* 1932; **25**: 1153–86

23 Allison AC. Protection afforded by sickle-cell trait against subtertain malarial infection. *BMJ* 1954; i: 290–4

24 Allison AC. Polymorphism and natural selection in human populations. *Cold Spring Harb Symp Quant Biol* 1964; **29**: 137–49

25 Ruwende C, Khoo SC, Snow RW *et al.* Natural selection of hemi- and heterozygotes for G6PD deficiency in Africa by resistance to severe malaria. *Nature* 1995; **376**: 246–9

26 Willcox M, Bjorkman A, Brohult J, Pehrson PO, Rombo L, Bengtsson E. A case-control study in northern Liberia of *Plasmodium falciparum* malaria in haemoglobin S and beta-thalassaemia traits. *Ann Trop Med Parasitol* 1983; **77**: 239–46

27 Allen SJ, O'Donnell A, Alexander ND *et al.* Alpha+-Thalassemia protects children against disease caused by other infections as well as malaria. *Proc Natl Acad Sci USA* 1997; **94**: 14736–41

28 Williams TN, Maitland K, Bennett S *et al.* High incidence of malaria in alpha-thalassaemic children. *Nature* 1996; **383**: 522–5

29 Modiano D, Petrarca V, Sirima BS *et al.* Different response to *Plasmodium falciparum* malaria in West African sympatric ethnic groups. *Proc Natl Acad Sci USA* 1996; **93**: 13206–11

30 Hill AV, Allsopp CE, Kwiatkowski D *et al.* Common west African HLA antigens are associated with protection from severe malaria. *Nature* 1991; **352**: 595–600

31 Hill AVS, Yates SN, Allsopp CE *et al.* Human leukocyte antigens and natural selection by malaria. *Philos Trans R Soc Lond B Biol Sci* 1994; **346**: 379–85

32 Hill AVS, Elvin J, Willis AC *et al.* Molecular analysis of the association of HLA-B53 and resistance to severe malaria. *Nature* 1992; **360**: 434–9

33 Gilbert SC, Plebanski M, Gupta S *et al.* Association of malaria parasite population structure, HLA, and immunological antagonism. *Science* 1998; **279**: 1173–7

34 Fernandez-Reyes D, Craig AG, Kyes SA *et al.* A high frequency African coding polymorphism in the N-terminal domain of ICAM-1 predisposing to cerebral malaria in Kenya. *Hum Mol Genet* 1997; **6**: 1357–60

35 Fowke KR, Nagelkerke NJ, Kimani J *et al.* Resistance to HIV-1 infection among persistently seronegative prostitutes in Nairobi, Kenya. *Lancet* 1996; **348**: 1347–51

36 Scorza Smeraldi R, Fabio G, Lazzarin A, Eisera NB, Moroni M, Zanussi C. HLA-associated susceptibility to acquired immunodeficiency syndrome in Italian patients with human-immunodeficiency-virus infection. *Lancet* 1986; ii: 1187–9

37 Kaslow RA, Duquesnoy R, VanRaden M *et al.* A1, Cw7, B8, DR3 HLA antigen combination associated with rapid decline of T-helper lymphocytes in HIV-1 infection. A report from the Multicenter AIDS Cohort Study. *Lancet* 1990; **335**: 927–30

38 McNeil AJ, Yap PL, Gore SM *et al.* Association of HLA types A1-B8-DR3 and B27 with rapid and slow progression of HIV disease. *QJM* 1996; **89**: 177–85

39 Kaslow RA, Carrington M, Apple R *et al.* Influence of combinations of major histocompatibility genes on the course of HIV-1 infection. *Nat Med* 1996; **2**: 405–11

40 Kroner BL, Goedert JJ, Blattner WA, Wilson SE, Carrington MN, Mann DL. Concordance of human leukocyte antigen haplotype-sharing, CD4 decline and AIDS in hemophilic siblings. Multicenter Hemophilia Cohort and Hemophilia Growth and Development Studies. *AIDS* 1995; **9**: 275–80

41 Dean M, Carrington M, Winkler C *et al.* Genetic restriction of HIV-1 infection and progression to AIDS by a deletion allele of the CKR5 structural gene. Hemophilia Growth and Development Study, Multicenter AIDS Cohort Study, Multicenter Hemophilia Cohort Study, San Francisco City Cohort, ALIVE Study. *Science* 1996; **273**: 1856–62

42 Smith MW, Dean M, Carrington M *et al.* Contrasting genetic influence of CCR2 and CCR5 variants on HIV-1 infection and disease progression. Hemophilia Growth and Development Study (HGDS), Multicenter AIDS Cohort Study (MACS), Multicenter Hemophilia Cohort Study (MHCS), San Francisco City Cohort (SFCC), ALIVE Study. *Science* 1997; **277**: 959–65

43 Winkler C, Modi W, Smith MW *et al.* Genetic restriction of AIDS pathogenesis by an SDF-1 chemokine gene variant. ALIVE Study, Hemophilia Growth and Development Study (HGDS), Multicenter AIDS Cohort Study (MACS), Multicenter Hemophilia Cohort Study (MHCS), San Francisco City Cohort (SFCC). *Science* 1998; **279**: 389–93

44 Turner MW. Mannose-binding lectin: the pluripotent molecule of the innate immune system. *Immunol Today* 1996; **17**: 532–40

45 Deslys J-P, Jaeglyy A, d'Aignaux JH, Mouthon F, Billette de Villemeur T, Dormont D. Genotype at codon 129 and susceptibility to Creutzfeldt-Jacob disease. *Lancet* 1998; **351**: 1251

46 Brown P, Cervenakova L, Goldfarb LG *et al.* Iatrogenic Creutzfeldt-Jakob disease: an example of the interplay between ancient genes and modern medicine. *Neurology* 1994; **44**: 291–3

47 Marquet S, Abel L, Hillaire D *et al.* Genetic localization of a locus controlling the intensity of infection by *Schistosoma mansoni* on chromosome 5q31-q33. *Nat Genet* 1996; **14**: 181–4

48 Pier GB, Grout M, Zaidi T *et al.* *Salmonella typhi* uses CFTR to enter intestinal epithelial cells. *Nature* 1998; **393**: 79–82

49 Wilson AG, Symons JA, McDowell TL, McDevitt HO, Duff GW. Effects of a polymorphism in the human tumor necrosis factor alpha promoter on transcriptional activation. *Proc Natl Acad Sci USA* 1997; **94**: 3195–9

50 McGuire W, Hill AV, Allsopp CE, Greenwood BM, Kwiatkowski D. Variation in the TNF-alpha promoter region associated with susceptibility to cerebral malaria. *Nature* 1994; **371**: 508–10

51 Kwiatkowski D, Molyneux ME, Stephens S *et al.* Anti-TNF therapy inhibits fever in cerebral malaria. *QJM* 1993; **86**: 91–8

52 van Hensbroek MB, Palmer A, Onyiorah E **et al**. The effect of a monoclonal antibody to tumor necrosis factor on survival from childhood cerebral malaria. *J Infect Dis* 1996; **174**: 1091–7

53 Bellamy R, Ruwende C, McAdam KP *et al.* Mannose binding protein deficiency is not associated with malaria, hepatitis B carriage nor tuberculosis in Africans. *QJM* 1998; **91**: 13–8

54 Daniels SE, Bhattacharrya S, James A *et al.* A genome-wide search for quantitative trait loci underlying asthma. *Nature* 1996; **383**: 247–50

55 Davenport M, Hill AVS. Reverse immunogenetics: from HLA-disease associations to vaccine candidates. *Mol Med Today* 1996; **2**: 38–45

56 Davenport MP, Ho Shon IA, Hill AV. An empirical method for the prediction of T-cell epitopes. *Immunogenetics* 1995; **42**: 392–7

57 Hoffman SL, Franke ED. Inducing protective immune responses against the sporozoite and liver stages of *Plasmodium*. *Immunol Lett* 1994; **41**: 89–94

58 Schneider J, Gilbert SC, Blanchard TJ *et al.* Enhanced immunogenicity for CD8+ T cell induction and complete protective efficacy of malaria DNA vaccination by boosting with modified vaccinia virus Ankara. *Nat Med* 1998; **4**: 397–402

59 Hill AV. The immunogenetics of human infectious diseases. *Annu Rev Immunol* 1998; **16**: 593–617

60 Davenport MP, Quinn CL, Chicz RM *et al.* Naturally processed peptides from two disease-resistance-associated HLA-DR13 alleles show related sequence motifs and the effects of the dimorphism at position 86 of the HLA-DR beta chain. *Proc Natl Acad Sci USA* 1995; **92**: 6567–71

61 Kostrikis LG, Huang Y, Moore JP *et al.* A chemokine receptor CCR2 allele delays HIV-1 disease progression and is associated with a CCR5 promoter mutation. *Nat Med* 1998; **4**: 350–3

62 Risch N, Merikangas K. The future of genetic studies of complex human diseases. *Science* 1996; **273**: 1516–7

Communicating genetic risk information

Theresa M Marteau

Psychology and Genetics Research Group, King's College, London, UK

It is envisaged that genetic information will be used, together with other types of information, to assess individuals' risks of developing a variety of common conditions. Such risk assessments will involve providing probabilistic information partly based upon results of genetic tests in order to facilitate behaviour change without causing excessive anxiety. The behaviours targeted for change are likely to include adherence to prescribed medication, alteration to diet, increasing levels of exercise and quitting smoking. This paper reviews research already conducted on perceptions of risk and genes, methods of facilitating behaviour change and reducing anxiety following various types of risk assessment. Although risk assessment and interventions to reduce risks have been conducted for over 20 years, very little rigorous research exists. For the investments in the new genetics to be realised, research is now needed both in how individuals respond to risk information that involves the use of genetic information and in how to facilitate and maintain behaviour change to reduce such risks.

How best to communicate genetic information depends in part upon the objectives of providing such information. Where the information predicts the onset of an untreatable condition, such as Huntington's disease, the objective of providing such information is a good psychological adjustment: relief from uncertainty, planning for the onset of a disease, and avoidance of depression and suicide. Similarly, when genetic tests are used to diagnose fetal abnormalities, with the offer of termination of affected pregnancies, optimal outcomes are defined in terms of informed decision-making and psychological adjustment of the parents. Achieving these outcomes depends upon people receiving information and emotional support, traditionally labelled as 'genetic counselling', to help them make decisions about genetic testing and to adjust to their test results.

In the next decade, genetic information will be used in the diagnosis, treatment and prediction of common conditions, such as hypertension, diabetes and different cancers[1]. A new taxonomy of diseases is being developed, based on genotype as opposed to phenotype, resulting in more effective and efficient development and prescribing of drugs. Genetic information will also be used as one of several risk factors in predicting disease prior to intervening to reduce risks. As the uses of genetic

Correspondence to: Prof. Theresa M Marteau, Psychology and Genetics Research Group, Level 5, Thomas Guy House, GKT Medical School, Guy's Campus, London SE1 9RT, UK e-mail: theresa.marteau@kcl.ac.uk

information expand to include diagnosis, treatment and prediction of treatable conditions so the aims of providing such information will expand beyond psychological adjustment to include understanding of gene and environment interactions and risk, as well as facilitating behaviour change to reduce risks. This expanding array of objectives makes it clear that counselling is only one of the many skills needed when communicating genetic information about risks of common, treatable conditions.

Introducing genetics into the diagnosis, treatment and prediction of common diseases raises the question of what individuals should be told and how such information is best conveyed. There is already some evidence concerning effective ways of communicating diagnoses[2], treatment[3] and predictions of illness[4]. While little of this evidence has involved genetic information, it seems likely that many of the basic principles of communication hold good regardless of the type of information being used. Research over the next decade will be needed to investigate whether this is the case.

The focus of this paper is upon the psychological effects arising from the use of genetic testing to predict common conditions. Such predictions will involve providing probabilistic information, partly based upon the results of genetic tests, in order to facilitate behaviour change. This needs to be achieved without causing excessive anxiety. Literature relevant to each of these four aspects of predicting disease are reviewed as a starting point for the development and evaluation of effective ways of communicating genetic information in this context.

Predicting responses to genetic information

Responses to any information are influenced not only by the information but also by an individual's pre-exiting perceptions. Individual's perceptions of risk and of genes form the basis for their responses to genetic information. Understanding these representations and how they interact with information about diagnosis, treatment and prediction provide the starting point for developing methods of communicating genetic information that result in the information being understood and which facilitate individuals acting to reduce risks, without experiencing excessive levels of anxiety.

Perceptions of genes

Beliefs that some illnesses, as well as an individual's constitution or vulnerability to illness 'run in families' are well rooted in Western cultures[5,6]. Heredity, for example, was seen as the major determinant of health in a sample of the general population in France[7]. In a UK study, heredity was seen as the third major cause of illness, after germs and lifestyle[8]. Recent studies suggest that the public see many illnesses, character traits and behaviour as multifactorial in origin, encompassing genetic as well as environmental causes[9]. A systematic review of 49 studies reporting causal beliefs for heart disease showed that the mean number of causes given was six, with the three most frequently provided being lifestyle (reported in all 49 studies), chronic stress (reported in 44/49 studies) and heredity (reported in 39/49 studies)[10]. What these studies do not tell us is how people's causal beliefs interact. Thus, while most people perceive stress, heredity and lifestyle to play a role in the development of heart disease, how do they see the relationships between these causes?

Beliefs about the causes of a disease are important in that they affect beliefs about its controllability, thereby affecting emotional adjustment and motivation to engage in behaviour that might reduce risks[9]. It is often assumed, erroneously, that providing individuals with information about risks to their health motivates them to change their behaviour to reduce such risks. It has been suggested that providing risk information based upon an analysis of DNA will also motivate people to engage in preventive behaviours[11]. There is, however, preliminary evidence to suggest that emphasising genetic predispositions to illness may actually de-motivate people from engaging in risk-reducing behaviours by making them think that the condition is neither preventable nor controllable. In an analogue study, students were asked to imagine that they had received a positive screening test result showing that they were at increased risk for heart disease or arthritis[12]. For half the participants, the screening test was described as a new genetic test which had detected a dominant gene for the condition. For the other half, the screening test was described only as a new test which had detected as increased risk for developing the condition. When risk to either disease was determined by a genetic as opposed to an unspecified test, the condition was seen as less preventable. Similarly, a recent study found that informing smokers of a genetic predisposition to lung cancer did not make them more likely to stop smoking, although it did make them more fearful[13].

Perceiving an illness as genetically-determined seems also to result in the condition being seen as more serious, possibly due to the perceived uncontrollability of genetically-determined conditions. Parents whose children had been identified as being at increased risk for familial hypercholesterolaemia following neonatal screening were interviewed about their experiences of the screening programme[14]. Responses to screening seemed to vary according to perceptions of the underlying cause of the positive screening test result. When parents perceived the test as detecting raised cholesterol, the condition was perceived as familiar, dietary in origin, controllable and not very threatening. When the test was seen as detecting a genetic problem, the condition was perceived as uncontrollable and hence more threatening, as illustrated by parents of two children who were identified as being at increased risk for FH:

Heredity levels, as I understand it, cannot be affected by diet. So, you know, I feel as though this death sentence has been put on my little boy.

It's the word heredity that sets all the alarm bells ringing – it sort of suggests that there is nothing you can do about it.

The extent to which these findings are attributable to risk assessments based upon genetic as opposed to other risk factors remains to be determined. A study is now underway in which individuals from families affected by familial hypercholesterolaemia are being randomly allocated to receive one of two risk assessments: one which incorporates genetic testing, the other of which does not (further details available from the author).

Research is needed to determine the content and structure of people's causal beliefs about common conditions. This will serve as a basis for predicting responses to risk assessments that incorporate genetic testing. This will also provide the basis for developing ways of presenting the results of such risk assessments to minimise threat and to motivate individuals to reduce risks.

Perceptions of risk

Risk perception comprises two key components: (i) the likelihood of an adverse event; and (ii) the perception of the seriousness of the event itself. Given that research has concentrated almost exclusively upon the first of these components, the focus here is upon the likelihood of an adverse event. It remains to be determined the extent to which perceptions of seriousness contribute to perceptions of risk.

Individuals' perceptions of risk affect their decisions about whether to undergo risk assessment as well as the likelihood that they engage in risk reducing behaviours[15]. Perceptions of risk are affected by the information an individual is given, as well as how individuals process probabilistic and threatening information.

Perceptions of risk and objective risk, although related to each other, are not synonymous. The extent to which they deviate and the reasons for this have been the subject of extensive research. The explanations can be broadly categorised as ones involving cognitive processes, and those involving emotional ones.

In making judgements about the likelihood of an event, people use a number of decision rules or heuristics not only because demands of daily life force us all to become 'cognitive misers' but because the rapid answers provided by heuristics are often right[16]. Three of the more common ones are:

- Availability (assessment of the probability of an event based upon the ease with which instances of the event come to mind).
- Representativeness (assessment of the probability of an event by judging the degree to which that event corresponds to an instance of that category).
- Anchoring and adjustment (a general judgement process in which an initially generated or given response serves as an anchor and other information is used to adjust that response).

The tendency for people to overestimate the likelihood of rare events, such as plane crashes, and underestimate the likelihood of more common events, such as road traffic accidents can be explained by the availability heuristic[17]. The rarity of plane crashes makes them newsworthy; their wide reporting makes instances easy to call to mind and hence to be judged more likely than they are. In judging the likelihood that someone may suffer from heart disease, an individual will judge how representative they seem to be of the perceived class of such individuals (*e.g.* overweight, under chronic stress), to the neglect of base rate information. The use of the anchoring heuristic is a likely explanation for the results of the studies reported below from Klein[18] and Yamagishi[19].

In addition to using cognitive heuristics when estimating personal risks of adversity, people use other psychological processes in order to reduce the degree of threat the risk invokes. These processes result in risks being minimised. There is, for example, a well described tendency for all of us to think that misfortune is more likely to happen to others than to ourselves[20]. When misfortune does strike, a common response is to minimise the seriousness of the event[21]. So, for example, those found to have raised levels of cholesterol, rate having a raised cholesterol as significantly less serious than do those with levels in the normal range[22].

Risk information can be presented in many ways. Two key dimensions along which the presentation of risk information may vary are the quantification of the risk and the framing of the outcome.

Quantification of risk

Risks are presented to patients in a variety of ways: as relative and absolute risks, numerical probabilities and proportions, and using a large number of verbal descriptors[23–25]. Risks presented as relative rather than as absolute risks tend to have a greater impact. For example, doctors are more willing to prescribe drugs if evidence of their effectiveness is presented in terms of relative as opposed to absolute risks[26,27]. One experiment involving university students showed that emotional reactions to hypothetical risk factor test results were sensitive to different levels of relative, but not absolute, risk[18]. Their judgements about safety were similarly influenced more by relative than by absolute risks. Students were asked to imagine that they had either a 30% chance or a 60% chance of causing an accident, and that this was either 20% higher or 20% lower than the average, same-age, same-sex person. Those told that their risk was lower than average rated themselves as safer drivers. Participants said to have a 60% chance of causing an accident when the average was 80% rated themselves as safer drivers than those told their chance of having an accident was 30% when the average was 10%.

Judgements of riskiness are also affected by the number of adverse outcomes presented, irrespective of the total number of individuals affected. In a series of experiments in which students were asked to judge the perceived riskiness of a number of events. Cancer was perceived[19] as riskier when it was described as 'kills 1286 out of 10,000 people' than as 'kills 24.14 out of 100 people'. Similar results were obtained for the other 10 causes of death. Other studies have also found that risk perceptions are influenced by the use of large numbers[28,29].

Little is known about the consequences of using numerical as opposed to verbal expressions of probability, such as 'low risk' or 'high chance'. In a survey of antenatal clinics in England and Wales, the two most common ways of presenting negative or low risk test results to pregnant women who had undergone serum screening for risk of Down syndrome were as a verbal phrase (*e.g.* low risk, lower risk), reported by 43% of clinics, or as a verbal phrase together with a numerical probability, reported by 40% of clinics[25]. The consequences of presenting risks in these different ways are unknown.

People prefer to receive information about probabilities of events in numerical form but prefer to express the probabilities of events to others using words, such as doubtful or likely[30,31]. While some studies have found that providing numerical probabilities leads to better decision-making than the presentation of verbal probabilities (such as improbable, possible, almost certain)[30], other studies find no difference in decision-making according to whether one or other type of probabilistic information is provided[31,32].

One of the aims of presenting risks in clinical contexts is to convey a likelihood of an adverse event, avoiding certainty either that an event will not happen (false re-assurance) or that it definitely will (fatalism), without causing undue anxiety. The relative merits of numerical and verbal probabilities in achieving these outcomes is currently being investigated in a study in which those undergoing prenatal screening for Down syndrome are randomised to receive results either using verbal or numerical probabilities (further details available from the author).

Framing of the outcome

While acknowledging the importance of probabilistic information, prospect theory states that, in making decisions, people begin by determining whether they stand to win or lose according to some reference point[33]. Different presentations of factually equivalent information may change the reference point. Thus, when asked to state whether they would consider surgery given various probabilities of success for their hypothetical lung cancer, students, patients and physicians are more likely to choose surgery when the possible outcome is expressed as the probability of surviving as opposed to the probability of dying[34,35].

Much of this research on perceptions of risk has been conducted in laboratory settings involving judgements of events that are neither related to health nor personally threatening[36]. Mood affects risk perception and behaviour in the face of risks[37]. It remains to be determined how the negative mood states that often follow from informing individuals of increased risks to their health (see below) alter the processing and hence impact of risk information presented in different ways.

Facilitating behaviour change

Reducing risks identified from a risk assessment will very often require an individual to do something, such as take medication, alter their diet, increase their level of exercise or stop smoking. There is now a wealth of literature, including systematic reviews, evaluating interventions aimed at changing these behaviours using a range of methods in a variety of populations.

Multiple risk factor interventions

Many primary prevention programmes have been set up over the past 20 years with the aim of preventing heart disease through modifying cardiovascular risk factors. The interventions have mainly involved provision of information about risk and how to reduce this. The results of a systematic review and meta-analysis of 14 randomised controlled trials evaluating the effectiveness of more than one of several interventions aimed at stopping smoking, taking exercise, altering diet and taking medication to control hypertension and serum cholesterol, revealed modest changes in risk factors and no significant effects upon mortality[38]. The authors conclude that efforts to reduce risks should be focused upon high risk groups and that health protection through fiscal and legislative measures may be more effective. More benefit may be gained from interventions targeted at just one risk factor.

Medication adherence

Low adherence to prescribed, self-administered medication is very common. In a systematic review of 13 randomised trials of interventions to increase adherence, Haynes *et al*[39] found that while adherence could be increased by certain, usually complex, interventions including various combinations of information, counselling, reminders, self-monitoring, re-inforcement and family therapy, research effort needed to be directed towards developing and evaluating new approaches to adherence so that the benefits of medication can be realised.

It may be instructive to consider the ratio of investments made in developing new medicines to the investments made in developing new approaches to ensuring that the medication is taken and hence its benefits realised. Haynes *et al*[39] comment that, with the astonishing advances in medical therapeutics during the past two decades, it is remarkable that only a handful of rigorous trials of adherence interventions have been conducted, providing little evidence that medication adherence can be improved consistently within the resources usually available in clinical settings.

Dietary change

Two recent meta-analyses of randomised controlled trials of dietary behaviour interventions show that individual dietary interventions in primary care achieve modest improvements in diet and cardiovascular risk status that are maintained for up to 18 months[40,41]. While one review reports that the greater the intensity of the intervention, the greater the fall in blood cholesterol levels[41], authors of the other review commented that a low-intensity approach may be as effective as one involving considerable resources[40]. Davey *et al*[42], commenting on these two reviews, consider that their most striking finding is the great variation seen in the effects produced in the different studies. There are, however, insufficient trials using similar approaches to allow sub-analyses to identify the characteristics of more successful interventions. Much might be gained by reviewing the effectiveness of behavioural, cognitive and environmental interventions effective in other contexts as a first step in developing further interventions to alter people's eating habits.

Tang *et al*[41] conclude that more research is needed to develop better methods of communicating dietary advice and maintaining adherence to such advice. Whether the changes achieved by individual dietary advice can more effectively and efficiently be achieved by population-based interventions remains a key question.

Increasing exercise

Activity levels in sedentary populations are most effectively increased by encouraging people to take brisk walks as opposed to more elaborate forms of exercise such as working out in gymnasia or playing squash. In a systematic review of randomised controlled trials of interventions to increase exercise in healthy, free living adults, just 11 trials were identified, none conducted in the UK[43]. Interventions that encouraged walking and did not require people to attend a facility were most likely to lead to sustainable increases in overall physical activity.

Quitting smoking

Law and Tang[44] reviewed 188 randomised controlled trials of interventions designed to help people stop smoking. They found that advice and encouragement to stop smoking from a physician was as effective as behaviour modification techniques (including relaxation and the use of rewards and punishments), with about 2% stopping smoking. Advice and encouragement was particularly effective with high risk smokers including pregnant women (8% effective) and those with heart disease. Nicotine replacement therapy was effective in approximately 13% of smokers who sought help in stopping.

These systematic reviews of interventions to change behaviour make sobering reading. First it is apparent that, apart from smoking, very few rigorous trials of interventions have been conducted. From those that have been conducted, interventions have varying degrees of success, taken as a whole it is at best modest. The research portfolio is this area is thin, both theoretically and methodologically. It is, therefore, too soon to conclude that behaviour change is too difficult and costly.

Minimizing anxiety

Receiving information about increased risks of illness is associated with increases in anxiety[4]. How much anxiety an individual experiences depends upon how the information is presented as well as an individual's pre-existing level of anxiety and coping style[45]. It is to be expected that individuals will react with concern and anxiety when informed that they are at increased risk of developing a disease. There is some evidence to suggest that a certain amount of fear facilitates behaviours aimed at reducing risks of disease, such as altering diet[46,47]. By contrast, anxiety interferes with the ability to understand information[48] and seems to inhibit behaviours that might lead to the detection of disease such as attending for clinical examinations or screening. Kash and colleagues[49], for example, found that the higher women's levels of anxiety, the less likely they were to examine their own breasts or to attend for clinical examinations.

Several interventions have proved successful in reducing anxiety levels in those undergoing risk assessment, involving the provision of information and training in coping skills.

Information

Providing information to people before they undergo medical or surgical procedures reduces anxiety[50]. The type of information presented is important in achieving this effect. The information needs to be sufficiently simple so that the majority of those receiving it can read and understand it and foster a sense of control over outcomes. The little research there has been on the effects of providing different types of written information given to those undergoing risk assessments, suggests that significant reductions in anxiety can be made using this relatively inexpensive intervention[51,52].

Enhancing coping skills

Providing cognitively-based interventions before risk assessment and after test results has been shown to be effective in reducing negative mood states in those undergoing HIV testing. Those undergoing HIV testing who received cognitive behavioural stress management were significantly less depressed after testing than those who received standard counselling[53]. Those who tested positive for HIV and who received standard counselling plus stress prevention training were significantly less anxious than those receiving standard counselling[54]. Stress prevention training (SPT) involved six weekly, 60 min individual sessions based on stress inoculation training[55] and cognitive-behavioural treatments for anxiety and depression[56]. The intervention proved acceptable to the HIV-infected cohort. In addition, relatively inexperienced psychology students were able to learn the techniques quickly and administer SPT with acceptable quality and adherence. It is possible that fewer sessions of SPT would be effective in risk assessment for conditions that are more treatable than HIV seropositivity.

Counselling

There are few studies of the type of counselling provided to those undergoing predictive genetic testing for uncommon conditions such as Huntington's disease, inherited forms of breast and bowel cancer. While it seems highly likely that the pre and post test counselling provided has been important in contributing to the low levels of psychological distress reported in these undergoing such testing, there is no evidence concerning how important it has been and what have been the effective components. Lerman and colleagues[57] evaluated the impact of individualised breast cancer risk counselling on distress about breast cancer as well as general distress in women with a family history of the disease. Compared with women who received general health education, those who received the breast cancer risk counselling had lower levels of distress about breast cancer. Their general levels of distress were unaffected. Perhaps the most important finding from this trial was that women with lower levels of education showed the greatest reductions in distress. As in any trial, it is important to conduct subgroup analyses to determine those most in need of an intervention and amongst those, those who benefit most. This allows more efficient targeting of resources as well as avoiding what some consider a patronising policy that insists on 'counselling for all'.

Providing a risk assessment service

It is clear from the preceding section that much research needs to be conducted to determine how best to present risk information and facilitate behaviour to reduce risks of disease. Before the results of such research are available, there are several models of population screening that can serve as the basis for risk assessments involving genetic testing. These include population-based screening for cervical and breast cancer and cardiovascular disease. The section below refers very briefly to just some of the issues that need to be considered when setting up these new risk assessment programmes.

The team

A variety of media are used to communicate risk information, including oral, written and video images presented in interactive systems. While geneticists will be part of the teams running risk assessment programmes they are unlikely to be needed to present risk assessments for common conditions to patients. Other members of the team designing risk assessments will include experts in the disease for which risk is being assessed, as well as behavioural scientists. Actual risk assessments may most effectively be delivered by individuals trained in risk assessment and behaviour change.

The training

It is essential but frequently neglected, for those presenting any form of risk assessment to be informed themselves about the meaning of the test and any possible test result. Insufficient knowledge about tests they are offering has been documented amongst midwives and obstetricians presenting screening for Down syndrome to pregnant women[58], and amongst doctors and nurses working in oncology presenting predictive genetic testing for certain cancers[59]. Telephone interviews with physicians in the US who had ordered genetic testing for FAP showed that over 30% incorrectly interpreted test results[60].

The messages

The information presented can affect interest in risk assessment and motivation to change behaviour. This raises the question of how far people should be encouraged to undergo risk assessment and to change their behaviour. There is a consensus in the genetics community that genetic information concerning prediction of conditions for which there is no treatment or for which termination is the main option should be presented in a non-directive manner[61-62], with no attempt by health professionals to influence decisions. Non-directiveness, however, is rarely defined or operationalised. Leaving aside the problems of defining non-directiveness[63,64], it is generally agreed that it is clinically appropriate to be more directive when presenting predictive tests for conditions for which there is effective treatment, such as familial adenomatous polyposis. Quite how directive is a matter of debate that cannot yet be informed by data.

Motivations of those running risk assessment programmes may affect the information that is presented, those motivated by financial profit perhaps presenting the benefits of their risk assessment in a more positive light than those not so motivated. A content analysis of 28 leaflets about CF carrier testing from commercial and non-commercial organisations in the US and the UK showed that leaflets from commercial organisations, as opposed to non-commercial ones, contained information more likely to lead to undergoing the test, *e.g.* less positive information was provided about life with cystic fibrosis, and the option of abortion was never mentioned[65]. While information that is presented will need to be balanced, how should balance be judged: by opinions of health professionals, the public or by research on the consequences of providing the information, or perhaps all three?

Concluding comment

Expectations of the therapeutic gains from the new genetics are high[1]. Realising these therapeutic gains depends upon how people behave when presented with genetic-related risk assessments. How can people at identified risk of disease be helped to alter their life style? How can adherence to prescribed medication be enhanced? Answering these questions is fundamental to the success of much that the new genetics has to offer. As this review shows, we are some way from being able to do this. As the mysteries of DNA are revealed, it is time to invest more in revealing the mysteries of people's behaviour when confronted with risks to their health.

Acknowledgement

Theresa M Marteau is supported by The Wellcome Trust.

References

1 Bell J. The new genetics in clinical practice. *BMJ* 1998; **316**: 618–20
2 Fallowfield LJ. Giving sad and bad news. *Lancet* 1993; **341**: 476–8
3 Ley P. *Communicating with Patients: improving communication, satisfaction and compliance.* London: Croom Helm, 1988
4 Shaw C, Abrams K, Marteau T. Psychological impact of predicting individuals' risk of illness: a systematic review. *Soc Sci Med* 1999; (In press)
5 Blaxter M, Paterson E. *Mothers and Daughters: a three generational study of health attitudes and behaviour*. London: Heinemann, 1982
6 Davison C, Frankel S, Davey Smith G. Inheriting heart trouble: the relevance of common-sense ideas to preventive measures. *Health Educ Res* 1989; **4**: 329–40
7 Herzlich C. *Health and Illness: a social psychological analysis*. London: Academic Press, 1973
8 Pill R, Stott N. Concepts of illness causation and responsibility: some preliminary data from a sample of working class mothers. In: Currer C, Stacey M, eds. *Concepts of Health, Illness and Disease: a comparative approach*. Leamington Spar: Berg, 1986; 257–77
9 Marteau TM, Senior V. Illness representations after the human genome project: the perceived role of genes in causing illness. In: Petrie KJ, Weinman JA, eds. *Perceptions of Illness and Treatment: Current Psychological Research and Implications*. Amsterdam: Harwood Academic, 1997; 241–66
10 French D, Senior V, Weinman JA, Marteau TM. Causal attributions for heart disease: a systematic review. (submitted)
11 Farmer A, Owen MJ. Genomics: the next psychiatric revolution? *Br J Psychiatry* 1996; **169**: 135–8
12 Senior V, Marteau TM, Weinman J. Impact of genetic testing on causal models of heart disease and arthritis: analogue studies. *Psychol Health* 1999; (In press)
13 Lerman C, Gold K, Audrain J *et al.* Incorporating biomarkers of exposure and genetic susceptibility into smoking cessation treatment: effects on smoking-related cognitions, emotions, and behavior change. *Health Psychol* 1997: **16**: 87–99
14 Senior V, Marteau TM, Peters TJ. Will genetic testing for predisposition for disease result in fatalism? A qualitative study of parents responses to neonatal screening for familial hypercholesterolaemia. *Soc Sci Med* 1999; (In press)
15 van der Pligt J. Perceived risk and vulnerability as predictors of precautionary behaviour. *Br J Health Psychol* 1998; **3**: 1–14
16 Fiske ST, Taylor SF. *Social Cognition,* 2nd edn. New York: McGraw-Hill, 1991
17 Fischhoff B, Lichtenstein S, Slovic P, Derby SL, Keeney RL. *Acceptable Risk*. Cambridge: Cambridge University Press, 1981
18 Klein WM. Objective standards are not enough: affective, self-evaluative, and behavioral responses to social comparison information. *J Pers Soc Psychol* 1997; **72**: 763–74

19 Yamagishi K. When a 12.86% mortality is more dangerous than 24.14%: implications for risk communication. *Appl Cognit Psychol* 1997; **11**: 485–94
20 Weinstein ND. Why it won't happen to me: perceptions of risk factors and susceptibility. *Health Psychol* 1984; **3**: 431–57
21 Croyle RT, Yi-Chun Sun, Hart M. Processing risk factor information: defensive biases in health-related judgements and memory. In: Petrie KJ, Weinman JA, eds. *Perceptions of Health and Illness*. Amsterdam: Harwood Academic, 1997; 267–90
22 Croyle RT, Yi-Chun Sun, Louie DH. Psychological minimization of cholesterol test results: moderators of appraisal in college students and community residents. *Health Psychol* 1993; **12**: 503–7
23 Hallowell N, Statham H, Murton F, Green J, Richards M. 'Talking about chance': the presentation of risk information during genetic counselling for breast and ovarian cancer. *J Genet Counsel* 1997: **6**: 269–86
24 Marteau TM, Plenicar M, Kidd J. Obstetricians presenting amniocentesis to pregnant women: practice observed. *J Reprod Inf Psychol* 1993; **11**: 3–10
25 Allanson A, Michie S, Marteau TM. Presentation of screen negative results on serum screening for Down syndrome: variations across Britain. *J Med Screen* 1997; **4**: 21–2
26 Bobbio M, Demichelis B, Guistetto G. Completeness of reporting trial results: effects on physicians willingness to prescribe. *Lancet* 1994; **343**: 1209–11
27 Bucher HC, Weinbacher M, Gyr K. Influence of method of reporting study results on decision of physicians to prescribe drugs to lower cholesterol concentration. *BMJ* 1994; **309**: 761–4
28 Denes-Raj V, Epstein S. Conflict between intuitive and rational processing: when people behave against their better judgement. *J Pers Soc Psychol* 1994; **66**: 819-29
29 McFarland C, Miller DT. The framing of relative performance feedback: seeing the class as half empty or half full. *J Person Soc Psychol* 1994: **66**: 1061–73
30 Budescu DV, Weinberg S, Wallsten TS. Decisions based on numerically and verbally expressed uncertainties. *J Exp Psychol Hum Percept Perform* 1988; **14**: 281–94
31 Erev I, Cohen BL. Verbal versus numerical probabilities: efficiency, biases, and the preference paradox. *Organizational Behavior and Human Decision Processes* 1990; **45**: 1–18
32 Gonzalez-Vallejo CC, Erev I, Wallsten TS. Do decision quality and preference order depend on whether probabilities are verbal or numerical? *Am J Psychol* 1994; **107**: 157–72
33 Kahneman D, Tversky A. Prospect theory: an analysis of decision under risk. *Econometrica* 1979: **47**: 263–91
34 McNeil BJ, Pauker SG, Sox HC, Tversky A. On the elicitation of preferences for alternative therapies. *N Engl J Med* 1982; **306**: 1259–62
35 Wilson DK, Kaplan RM, Schneiderman L. Framing of decisions and selection of alternatives in health care. Paper presented at the *7th Annual Meeting of the Society of Behavioral Medicine*, March 1986, San Francisco
36 Brun W. Risk perception: main issues, approaches and findings. In: Wright G, Ayton P, eds. *Subjective Probability*. Chichester: John Wiley, 1994; 295–320
37 Pezza Leith P, Baumeister RF. Why do bad moods increase self-defeating behavior? Emotion, risk taking, and self-regulation. *J Pers Soc Psychol* 1996; **71**: 1250–67
38 Ebrahim S, Davey Smith G. Systematic review of randomised controlled trials of multiple risk factor interventions for preventing coronary heart disease. *BMJ* 1997; **314**: 1666–74
39 Haynes RB, McKibbon KA, Kanani R. Systematic review of randomised trials of interventions to assist patients to follow prescriptions for medicines. *Lancet* 1996; **348**: 383–6
40 Brunner E, White I, Thorogood M *et al.* Can dietary interventions change diet and cardiovascular risk factors? A meta-analysis of randomized controlled trials? *Am J Public Health* 1997; **87**: 1415–22
41 Tang JL, Armitage JM, Lancaster T, Silagy CA, Fowler GH, Neil HAW. Systematic review of dietary intervention trials to lower blood total cholesterol in free-living subjects. *BMJ* 1998: **316**: 1213–9
42 Davey Smith G, Ebrahim S. Commentary: Dietary change, cholesterol reduction, and the public health – what does meta-analysis add? *BMJ* 1998; **316**: 1220
43 Hillsdon M, Thorogood M. A systematic review of physical activity promotion strategies. *Br J Sports Med* 1996; **30**: 84–9

44 Law M, Tang JL. An analysis of the effectiveness of interventions intended to help people stop smoking. *Arch Intern Med* 1995; **155**: 1933–41

45 Marteau TM, Croyle RT. Psychological responses to genetic testing. *BMJ* 1998; **316**: 693–6

46 Sutton SR. Fear-arousing communications: a critical examination of theory and research. In: Eiser JR, ed. *Social Psychology and Behavioral Medicine*. Chester: Wiley, 1982; 303–37

47 Millar MG, Millar K. The effects of anxiety on response times to disease detection and health promotion behaviors. *J Behav Med* 1996; **19**: 401–13

48 Lerman C, Croyle R. Genetic testing for cancer predisposition: behavioral science issues. *J Natl Cancer Inst* 1995; **17**: 63–6

49 Kash KM, Holland, JC, Halper MS, Miller DG. Psychological distress and surveillance behaviors of women with a family history of breast cancer. *J Natl Can Inst* 1992; **84**: 24–30

50 Johnston M, Vogele C. Benefits of psychological preparation for surgery: a meta-analysis. *Ann Behav Med* 1993; **4**: 245–56

51 Wilkinson C, Jones JK, McBride J. Anxiety caused by abnormal result of cervical smear test: a controlled trial. *BMJ* 1990; **300**: 440

52 Marteau TM, Kidd J, Cuddeford L. Reducing anxiety in women referred for colposcopy using an information booklet. *Br J Health Psychol* 1996; **1**: 181–9

53 Antoni MH, Baggett L, Ironson G *et al.* Cognitive-behavioral stress management intervention buffers distress responses and immunologic changes following notification of HIV-1 seropositivity. *J Consult Clin Psychol* 1991; **59**: 906–15

54 Perry S, Fishman B, Jacobsberg L, Young J, Frances A. Effectiveness of psychoeducational interventions in reducing emotional distress after human immunodeficiency virus antibody testing. *Arch Gen Psychiatry* 1991 ; **48**: 143–7

55 Meichenbaum D. *Stress Innoculation Training*. Elmsford, NY: Pergamon, 1985

56 Beck AT, Rush AJ, Shaw BF, Emery G. *Cognitive Therapy of Depression*. New York: Guilford, 1979

57 Lerman C, Schwartz MD, Miller SM, Daly M, Sands C, Rimer BK. A randomized trial of breast cancer risk counseling: interacting effects of counseling, educational level, and coping style. *Health Psychol* 1996; **15**: 75–83

58 Smith DK, Shaw RW, Slack J, Marteau TM. Training obstetricians and midwives to present screening tests: evaluation of two brief interventions. *Prenat Diagn* 1995; **15**: 317–24

59 Chorley W, MacDermot K. Who should talk to patients with cancer about genetics? *BMJ* 1997; **314**: 441

60 Giardiello FM, Brensinger JD, Petersen GM *et al.* The use and interpretation of commercial *APC* gene testing for familial adenomatous polyposis. *N Engl J Med* 1997; **336**: 823–7

61 Andrews LB, Fullarton JE, Holtzman NA, Motulsky AG, eds. *Assessing Genetic Risks: Implications for Health and Social Policy*. Washington DC: National Academy, 1994

62 Royal College of Physicians, *Prenatal Diagnosis and Genetic Screening: Community and Service Implications*. London: Report of the Royal College of Physicians, 1989

63 Clarke A. Is non-directive genetic counselling possible? *Lancet* 1991; **338**: 998–1001

64 Lippman A. Prenatal genetic testing and screening: constructing needs and reinforcing inequities. *Am J Law Med* 1991; **17**: 15–50

65 Loeben GL, Marteau TM, Wilfond BS. Mixed messages: Presentation of information in cystic fibrosis screening pamphlets. *Am J Hum Genet* 1998; **63**: 1181–9

Genomics: the implications for ethics and education

Sandy M Thomas

Nuffield Council on Bioethics, London, UK

Over the past 10 years, the identification of disease genes has been expanding rapidly. Those identified in the earlier part of the decade were largely achieved through positional cloning and the majority are for relatively rare disorders which involve single genes. As the Human Genome Mapping Project has progressed, the rate of gene discovery has increased substantially through the development of new DNA sequencing techniques and *in silico* approaches. The human genome will have been largely sequenced by ther Spring 2000. We can expect the identification of large numbers of susceptibility genes for common multifactorial polygenic diseases as well as genes which are associated with human behavioural traits. Some of these advances hold out the prospects of real progress in the diagnosis and treatment of a wide range of disorders. However, for many individuals, increased knowledge about their genes will present ethical dilemmas which are difficult to resolve. There are also wider ethical issues which concern the use of genetic information by insurers and employers and yet others which concern ownership and access. In this chapter, the main ethical issues raised by the impact of genomics on healthcare are discussed. The role of education in enabling individuals and health professionals to meet these challenges is also considered.

An ethical framework for genomics

The application of ethical principles to genetics aims to provide an ethical and humanistic perspective in a rapidly developing field where health and disease are characterised largely by scientific terminology. Molecular complexity revealed by current research should not be allowed to distract from the fact that the subjects under study are human beings and their values. An increasing tendency on the part of some researchers to view human beings as 'gene carriers' and similar attitudes risk undermining both moral responsibility and social solidarity[1]. Social solidarity involves distributing benefits and losses across society as a whole and expressing equality of respect for persons.

Correspondence to: Dr Sandy Thomas, Director, Nuffield Council on Bioethics, 28 Bedford Square, London WC1B 3EG, UK

British Medical Bulletin 1999;**55** (No. 2): 429–445

With regard to the search for and availability of genetic information, two ethical requirements may be regarded as basic[2]. These include the limitation of harm and suffering to humans and the maintenance of respect for human beings and human dignity. Limiting harm and suffering is demonstrated by seeking to cure, to care and not to injure and thereby to establish and maintain health systems that deliver effective, affordable and timely treatment. Respect for the person is expressed in action and procedures that give due consideration to personal autonomy and integrity[3]. This is reflected in the obligation of doctors and researchers to seek informed consent, to maintain confidentiality and to respect privacy.

A distinctive aspect of genetic disorders is that they affect not only individuals but groups of related individuals. Genetic information about one person may reveal certain or probalastic information about their relatives including any unborn children. Information about the wider family can be obtained by testing or treating only one member. This aspect of genetic disorders raises many difficult ethical questions and it is often difficult to decide how best to exercise the two basic principles referred to above. For example, where there is a dilemma about disclosure of genetic information to relatives, non-disclosure might be seen by some as a form of paternalism that denies them their own autonomy. Alternatively, genetic tests could be treated like other medical information and remain confidential to individuals. Genetic knowledge arising from the application of genomic information may challenge the accepted framework of medical confidentiality for individuals.

Experience of ethical issues raised by increased genetic knowledge has been largely based on relatively rare disorders caused by mutation in single genes. The ethical issues here are relatively well understood and guidelines for health professionals about informed consent, confidentiality and genetic counselling have been established by advisory bodies in most countries that offer genetic testing services. While this experience may provide valuable insights for the application of genetic information about complex diseases, developing a parallel ethical framework which enables health professionals and ordinary people to cope with large amounts of information about susceptibility genes for common diseases will be a major challenge. Further ethical dilemmas will be posed by the discovery of genes which are associated with behavioural traits rather than specific diseases. Although there as yet are few confirmed reports of such genes, there is a danger that knowledge about genes which may influence intelligence, alcoholism and even crime will be presented to the general public in an over-simplified way. There are also concerns that genetic information may be used inappropriately by external agencies, such as the insurance industry and employers[4]. The ethical concerns that may confront individuals, their families and physicians when genetic information is presented in a clinical context are considered in the next section.

Ethical concerns in genetic testing: individuals and families

Genetic counselling

Genetic counselling had been defined[3,5] as 'the process by which patients of relatives at risk of a disorder that may be hereditary are advised of the consequences of the disorder, the probability of developing or transmitting it and of the way that it might be prevented, avoided or ameliorated'. The ethical standards which genetic counselling must meet are already widely accepted[6–8]. Those who provide the service have defined responsibilities. These are to ensure that genetic counselling is voluntarily undertaken and that it provides assessable and accurate information about patterns of inheritance and the disorder. Confidentiality must also be ensured to those receiving counselling. If there are good reasons to share the information with other relatives, this should be explained. It is also important to emphasise that consent to counselling or genetic testing does not constitute consent to act on any advice, to take any reproductive decision or to terminate a pregnancy[9].

For some conditions, such as cystic fibrosis and Huntingdon's disease, it is possible to predict the risk of occurrence on the basis of family history very precisely. For example, the nature of mutations in the Huntingdon's gene on chromosome 4 mean that the disease is a dominantly inherited condition. If one parent has the Huntingdon's gene mutation, there is a 50% chance that this gene will be passed on to his or her child. A similar pattern is shown by early onset Alzheimer's disease.

The ideal of 'non-directiveness' in genetic counselling has been widely endorsed. In a recent report, the Working Party on Mental Disorder and Genetics of the Nuffield Council on Bioethics questioned the clarity and feasibility of non-directiveness as a universal attainable aim and noted the importance of enabling individuals to make their own informed decisions at each stage of the counselling process. The available evidence indicates that, in general, reproductive intentions are rarely altered by genetic counselling. Rather, the main outcome appears to be that couples feel supported in their initial intentions[10].

Genetic testing

Genetic tests have been developed for a relatively small number of diseases. However, predictive testing for single gene disorders such as

Huntingdon's disease has been in progress for almost 10 years in the UK. Direct gene-based testing has been in place since 1993[11]. One of the most important observations to emerge is that the number of people seeking testing for Huntingdon's disease is much lower than was initially envisaged. Prior to identification of the Huntingdon gene, research suggested that about 75% of those at risk of inheriting the mutation from a parent would seek testing. It was widely considered that advantages of resolving uncertainty and an improved basis for planning lives would make testing the preferred choice of family members. However, probably less than 10% of those with an affected parent have elected to have counselling and of these about two-thirds opt to be tested[12]. There are also early indications that very few members of families that carry early onset Alzheimer's disease seek to have a genetic predictive test[13]. An important finding from research on genetic testing is that reactions to the availability of tests are specific to particular disorders. Different uptake rates for testing and outcomes have been reported for different adult onset conditions[14]. These may depend not only on the perception of a particular disease but also on the certainty of onset, the options for prevention and treatment and the implications for health care and life insurance.

Consent and impaired capacity

The ethical principle of respect for the autonomy and dignity of human beings underlies the legal requirement for consent to be sought prior to any genetic counselling or testing of adults. The Law Commission has recommended that statutory force be applied to the existing common law presumption that an adult has full legal capacity unless it is shown that he or she does not[15]. The law requires that, in the determination of a patient's capacity to consent to a procedure or not, the doctor or psychiatrist must be satisfied that the patient possesses the capacity to make a choice, understands the nature of procedure in broad terms, its principal risks and benefits as well as the consequences of not undergoing the procedure[16]. It has been noted that even for individuals able to give consent, fully informed consent is an unattainable ideal[17]: 'the ethically significant requirement is not that consent be complete but that it be *genuine*'. For an adult person deemed mentally incompetent to make decisions, a doctor must act in that patient's 'best interest'.

The genetic testing of children

In the UK, children aged between 16 to 18 years may give valid consent to treatment as if they were adult, provided that they are otherwise competent. A child below the age of 16 years may also consent to medical treatment if they are also judged able to fully understand what is involved in the treatment or procedure[18]. General issues of testing children are viewed as comparable to those raised in the genetic testing of adults. However, when the testing of children is proposed for reasons other than for diagnosis or treatment, ethical difficulties may arise[19]. For example, parents may want a child to be tested for a genetic disorder to eliminate uncertainty even if there is no available treatment, while the child may have a different view. Adolescents may wish to be tested but their parents may object. These situations raise complex issues of benefit and possible harm and caution should therefore be exercised before a decision is left solely to the child[20].

With regard to diagnostic testing, only in a few cases, such as phenylketonuria, are effective interventions available for genetic disorders in children. In such instances, a genetic test may assist in the diagnosis and, therefore, be in the child's best interests. Predictive genetic testing in children is more controversial. Clinical geneticists have generally opposed the testing of children for adult onset conditions as this denies the child the chance of making his or her own decisions as an adult and could even lead to discrimination within the family[21]. Whatever the ethical arguments, unless the testing does not explicitly serve the best interest of the child it would not be permissible in law. Similarly, carrier testing denies the possibility of children making their own decisions at a later date.

Genetic information and reproductive decisions

The primary use of genetic information is to inform reproductive choice and this can be used prenuptially, preconceptionally or prenatally. Prenatal genetic testing may produce information which may influence a decision on whether or not to seek a abortion. Many parents will accept genetic testing and abortion when a test can reveal the presence of a severe, early onset disorder for which there is no treatment. However, these tests should not be tied to conditions which might influence an individual's decision whether to be tested or not. Such a condition might be the willingness to consider an abortion as a condition of prenatal testing. Some pregnant women might take the test in order to secure reassurance and may be inadequately prepared for an adverse result[22].

Individual choices about having children can affect gene frequencies in later populations. For example, those who take steps to avoid having children with a dominant mutation such as Huntingdon's disease will contribute to the mutations becoming less frequent in future generations. These reductions have been welcomed by ethical committees[23]. For recessively inherited conditions, however, carrier frequencies in populations will remain unaffected.

There have been concerns about the provision about genetic testing services for conditions that some people would regard as part of normal human variation rather than disabilities. For example, in the case of achondroplasia, disabled rights groups have argued that availability of genetic testing is intrinsically eugenic. Parents should make their own decisions about whether or not to proceed with a pregnancy in the light of a fetus being diagnosed as having a genetic disorder and should be supported accordingly. The Nuffield Council on Bioethics Working Party on Mental Disorders and Genetics concluded that the best safeguard against eugenic pressures is properly informed, freely given consent.

Confidentiality and disclosure

The confidentiality of genetic information is protected in several ways by common law, statute, professional codes of practice and contracts of employment. The duty of confidentiality is not absolute. Both the law and professional guidelines allow for exceptional circumstances when an individual cannot be persuaded to inform a relative who has a legitimate right to know, although debate continues. However, some relatives may not wish to be presented with genetic information they would rather not have. This is sometimes referred to as 'the right not to know'. Debate concerning disclosure to relatives who may be aware of an inherited disorder in the family is finely balanced. Some argue for the patient's right to absolute confidentiality while others consider that the doctor or genetic counsellor has a duty to disclose information in exceptional circumstances[24].

Complex diseases: susceptibility genes

Common complex diseases which are multifactorial and polygenic will present new challenges to the ethical framework outlined above. In broad terms, the issues will be much the same. However, because the genetic information will be probabilistic rather than relatively clear cut as in monogenic diseases, ethical issues may be harder to resolve. The example of the *apoE* gene and its role in Alzheimer's disease illustrates one kind of ethical difficulty. Although the apolipoprotein E4 allele is more common in patients with Alzheimer's disease than healthy controls, it only accounts for about 15% of susceptibility to the disease. The increased risk associated with apoE4 is small and, because it is calculated for the whole population, does not take into account individual genetic and environmental variation. Moreover, about half of all patients with Alzheimer's disease do not possess an E4 allele. Because a genetic test can indicate no more than a somewhat increased susceptibility, some studies have concluded that it is not appropriate for either diagnosis or prediction in individual members of the population as a whole[25], although debate on this continues[26]. Knowledge of a predisposition to a disorder may be useful even if the risk is low provided some medical or lifestyle change could reduce susceptibility to, or severity of, the disorder. On the other hand, the same knowledge might be harmful if there were no chances to reduce the susceptibility and it became a source of anxiety. Applying the ethical principle of limiting harm and suffering would be difficult to apply in these situations where disorders are associated with relatively slight genetic predispositions.

Showing respect for others in terms of their autonomy and integrity in the context of susceptibility genes may also present difficulties for the health professional. For example, doctors will need to determine whether, when and how to offer information concerning genetic tests for disease predispositions. They will need to advise patients about taking the test, decide how to disclose the information and explain the degree of risk whilst avoiding undue alarm. Many common diseases are probably influenced by variants in several genes with each individual allele having a relatively small effect. The fact that susceptibility genes are neither necessary nor sufficient to cause the disease, limits their use in genetic tests for either diagnosis or prediction[27]. Thus in familial hypercholesterolaemia (FH), diagnosis by the presence of a specific allele is of no better predictive value than a cholesterol assay. Furthermore, the development of heart disease may be strongly influenced by many additional factors such as diet and smoking and it has been pointed out that this makes the predictions of disease risk by genetic testing of FH impractical for most individuals[28].

It has been claimed that gene identification will be very valuable in personalising risks and that the increased precision in calculating individual risk will be of enormous clinical benefit[29]. Evidence to support these claims is lacking. What level of risk might susceptibility genes confer? The risk of developing schizophrenia if the patient has a first-degree relative is approximately 10%. Candidate susceptibility genes for schizophrenia are being identified and one of these is a variant in the serotonin receptor (5HT2a) gene which occurs in 70% of people with schizophrenia in the UK[30]. The risk to an individual who has the serotonin receptor variant and a sibling affected by schizophrenia of developing the disease is 12.3%, slightly higher than the 10% empiric risk figure. The Working Party on Mental Disorders and Genetics of the

Nuffield Council on Bioethics recently concluded that this level of increase is unlikely to be of much use in the clinic given that a reduction in uncertainty is one of the most common reasons cited for taking a DNA test[31]. The overall value of genetic counselling and testing in the common multifactorial polygenic diseases will depend largely on two criteria: (i) the ability to calculate individual risks; and (ii) the ability to identify preventive measures to reduce the risk of the disorder occurring in individuals at high risk[32]. It has been pointed out that even if sufficient susceptibility genes were identified to account for some 30% of variation in risk between different people in a population, absence of knowledge about interaction between genes and the environment would still make it problematic to predict individual risk accurately[33]. A further difficulty with the utility of genetic tests for polygenic multifactorial diseases is the limited knowledge about the number of susceptibility genes involved and how they combine and interact. However, research on these genes will certainly improve our understanding of causal mechanisms and lead to novel therapeutic or preventive strategies.

How will the identification of susceptibility genes for common complex diseases affect current ethical frameworks for research and chemical practice in genetics? Because the contribution to risk of any one susceptibility gene may be small, genetic counsellors will need to emphasise the limitations of scientific knowledge. Large scale testing for such genes would produce false positives and negatives which might burden NHS services accordingly. The Nuffield Council Working Party on Mental Disorders and Genetics recently recommended that genetic testing for susceptibility genes providing predictive or diagnostic input of certainty comparable to, or lower than, that offered by an apoE test for Alzheimer's disease should be discouraged unless and until the information can be put to effective and preventative or therapeutic use. Susceptibility genes may also influence more than one condition. For example, different alleles of *apoE* are also associated with heart disease. Where there is a possibility that additional information may be revealed by taking a genetic test, doctors have been advised to discuss this with the patient beforehand.

Direct marketing of genetic tests

The development of genomics has led to the suggestion that the small range of genetic test kits marketed directly to the public will increase rapidly over the next few years. The view has been put forward[34] that while 'susceptibility screening may be bad science, it is likely to be excellent business'. Screening tests applicable to the general population hold out promise of substantial profits for those corporations that can develop and patent tests and techniques ahead of their competitors'.

These commercial pressures might lead to the promotion of susceptibility testing despite the fact that it might not be advisable or appropriate. The present voluntary code of practice for directly marketed tests in the UK may prove unworkable in the longer term[35].

Wider uses of genetic information: insurance and employment

Concerns over the confidentiality of genetic information extend to the use of that information by outside agencies such as the insurance industry and employers. There has been wide debate in Europe and the US about the use of genetic information for insurance purposes. In the UK, attention has been largely directed towards life insurance. The actuarial significance of genetic information is complex but crucially important. Even in monogenic disorders such as Huntingdon's disease where there is a calculable and actuarially significant reduction in life expectancy, there is, nevertheless, significant variability between individuals which may be due to environmental or genetic factors. Information about susceptibility genes is likely to be of relatively limited actuarial use, revealing little or nothing about any one individual's level of risk. Genetic information which the insurance industry could use to calculate an increased risk to a given individual with specific susceptibility genes is not yet available.

The Association of British Insurers (ABI) has issued a code of practice[36] on the use and handling of genetic information by insurance companies. Although the code states that the applicants will not have to undergo genetic tests, they will, nevertheless, be asked to disclose existing relevant genetic test results. These will only be used if the results demonstrate a clearly increased risk of genetic disease. There is currently a UK moratorium whereby existing genetic test results need not be revealed for applications of life insurance up to the value of £100,000 linked to a mortgage on a primary place of residence. The Human UK Genetics Advisory Commission has proposed that a general moratorium on the disclosure of genetic test results should only be lifted when there is a clear demonstration of the actuarial relevance of specific sorts of genetic test for specific insurance products[37]. This suggestion has been rejected by the ABI which takes the view that there are already eight specific genetic tests which provide actuarially significant information[38]. It should be borne in mind that any exaggeration of the actuarial implications of genetic tests amounts to unfair discrimination. It has been suggested that a system which monitors whether UK insurers are discriminating unfairly on the basis of genetic test results needs to be put in place.

Employment

Employment is another area in which there is a danger of genetic discrimination. The application of genetic information for employment purposes has hitherto been less controversial than its use in insurance. It has been recommended by a UK ethical committee that 'genetic screening of employees for increased occupational risk ought only to be contemplated where:

1 There is strong evidence of a clear connection between the working environment and the development of the condition for which genetic screening can be conducted.

2 The condition in question is one which seriously endangers the employee or is one in which an affected employee is likely to present a serious danger to third parties.

3 The condition is one for which dangers cannot be eliminated or significantly reduced by reasonable measures taken by the employer to modify or respond to the environmental risks'[39]. Genetic screening of employees is extremely limited in the UK.

The Ministry of Defence screens potential applicants to the forces for the sickle cell trait where there is a risk that they would be exposed to atypical atmospheric pressures. Another example of a genetic test linked to occupational hazards concerns individuals exposed to organic solvents who have an increased risk of developing Goodpasture's disease if they have a particular genotype (HLA type DR2). Wider use of these kinds of tests by employers may raise important issues about discrimination. At present, there is little legal or regulatory framework to address these issues in the UK.

What are the ethical issues which might arise? Although employers in the UK and elsewhere are obliged to act within the framework of an Equal Opportunities policy, genetic test data could be used to achieve a healthier workforce by excluding potential employees with adverse test results. Such individuals might also be disadvantaged by exclusion from early retirement schemes operated by pension funds. The UK Disability Discrimination Act, which obliges employers to treat persons with a disability on equal terms with other employees (unless they can justify their behaviour), does not cover instances where genetic tests suggest that an individual will or may directly develop disability.

The House of Commons Science and Technology Select Committee suggested that a Privacy Bill should be enacted to make misuse of genetic information, both a criminal and civil offence, but this has yet to be taken up[40]. In Denmark, legislation on the use of personal health information in employment decisions has resulted in stringent regulations on the use of genetic data. However, it has been noted that enforcement of these regulations is relatively weak[41]. In the US, some states have prohibited the acquisition and/or use of genetic information about current or prospective employees.

Genes influencing behaviour

The mapping and sequencing of the human genome together with functional genomic analysis will provide a great deal of information about genes which are not directly implicated in disease. The widely used 'shorthand' for genetic traits such as schizophrenia, intelligence, criminality or even divorce by some scientists and the media is both inaccurate and unhelpful[42]. This approach suggests that a deterministic one-to-one relationship exists between a gene and its characteristic and between genotype and phenotype. Discussion about genes associated with human behaviour suggest that there may be a future demand from the public for tests, particularly from those who wish to optimise the characteristics of their unborn children. These scientific developments are at an early stage and many of the reports of genes associated with, for example, criminal behaviour remain unconfirmed. Research of this kind should be ideally linked to therapeutic improvement where the genetic information can be applied to benefit the patient. The real danger of identifying personality traits *per se* is that this information may be put to dubious use. Moreover, most of the behavioural traits which are identified will, in most cases, have small effects and interact with external factors. For example, two studies have demonstrated that the allele of the dopamine receptor DRD4 which has seven repeats is associated with significantly higher levels of the personality trait of novelty-seeking[43]. This is a normal variant or polymorphism, although the association between the allele and the trait is weak. Only 4% of the normal population variation was accounted for in association studies. More controversially, there is preliminary evidence that this allele is over-represented in people with attention deficit hyperactivity disorder (ADHD) and under-represented in people with major depression[44].

Ethics in genomics research

In a rapidly advancing area such as genomics, it is inevitable that research and clinical work will be closely associated. The diversity of positions on the control of DNA samples and information reflect the difficulty of interpreting human genetics research into mainstream medicine[45]. Research aimed at identifying genes associated with particular disorders may rely on access to DNA samples from large collections of families with the disorder. For rare disorders, researchers are frequently better placed than clinicians to provide information to patients. Indeed, some researchers have even set up specialised clinics to which family members at risk may be referred for genetic counselling[46]. Provided appropriate guidelines are observed, such arrangements are not likely to raise specific ethical problems. More complex ethical issues arise when researchers using DNA samples collected for research purposes discover information which is of clinical significance to the donor. When a disease-linked gene has been located and significant mutations found, the question then arises as to how to deal with the information. In addition, for those individuals who have been found to have the relevant mutation, there may be implications for relatives who have not consented to take part in the research. The ethical issue here concerns the fact that the informed consent required for research does not usually include consent for disclosure of patient-identifiable data to clinics outside the research environment. To provide an individual with research information about mutations which they might have and which could not have been anticipated could be to give them information they would prefer not to have and which they or others are not prepared for. There is also a risk that the quality control procedures used, for example in direct predictive testing in Huntingdon's disease, are unlikely to be used in research. As a general rule[47], it has been suggested that those individuals who consent to take part in research should be told that results from analysis of their DNA 'will not be given to them'.

Another ethical issue that may arise in research concerns access to established data banks by other researchers. Provided that the requirement is only for anonymised aggregate data, there are likely to be few problems. However, researchers may also wish to have access to named data or to seek use of genetic registers or cohorts as a means of identifying groups of individuals with rare conditions. To avoid future complications, the possibility of further research should be included in the original consent process. Recent guidance for researchers suggested that for some research it will be necessary to meet with the individual again to collect information or to take further samples. Consent for data sharing can then be sought in the normal way[48]. New research on existing samples should be submitted for approval to a research ethical committee before proceeding.

Genetic registers or data banks will contain genetic information which must be kept confidential. A frequent concern is that agencies – including the health and social services, the police, sectors of the criminal justice system, insurance and other finance agencies – may be able to have access to research information. To use research data without permission would need strong justification[49]. The European Human Rights Convention and recent EU initiatives on data protection address the protection of privacy. Ethical guidance recommends that participation in research which involves genetic investigation should not require participants to state that they have been tested for genetic conditions in insurance applications[50].

Implications for education

Although uncertainties remain about potential increases in demand for genetic counselling and testing, genetics will play a greatly expanded role in health care. It has been predicted that the increasing demand for genetic information and services could require a major reorganisation of genetic services[51]. Primary health teams in particular are likely to face increased demands for genetic information about genetic disorders from their patients[52]. As the Human Genome Project progresses and awareness grows within the general public about a range of disorders and predispositions that may be present in their families, there will be an inevitable increase in demand for advice. In the majority of cases, patients will not need to be referred to specialist services but will require assessment and reassurance. We have seen that many susceptibility genes are unlikely to increase an individual's risk to a degree that would merit specialist counselling, at least for the purpose of discussion about the possibility of genetic testing. Apart from a small number of people with rare, single gene disorders the need for specialist referral should be low. If this proves to be the case, it will be necessary to balance any inappropriate demand for specialist counselling against other health care priorities. There is, nevertheless, an ethical obligation to identify the few people who genuinely need specialist genetic counselling and to provide any useful information to those who do not[53].

It has also been predicted that the major investments in pharmacogenomics by pharmaceutical companies will lead to the development of genetic testing of patients for their ability to tolerate major drugs within five years. Furthermore, it appears probable that this kind of testing will become increasingly routine within the decade. There are obvious health benefits to the patient who is less likely to experience undesirable side effects from drugs that are not well tolerated. One implication of these developments is that genetic testing for these characteristics may need to become part of primary care. For example, if a particular patient is prescribed medication which is suited to one genotype rather than another, testing will need to be within the general practitioner setting. The provision of medication to patients is also likely to involve genetic testing of the patient.

Genetic services in the UK are presently organised into regional centres. Their main role has been to provide genetic counselling and testing services for relatively rare single gene disorders and congenital conditions. Much of the demand had been driven by reproductive issues, but predisposition for common diseases is likely to play an expanding role. This may lead to the need for a substantial increase in the availability of genetic expertise, although how this should be distributed between primary and secondary care is uncertain. Strategies to deal with the increased demand for information and advice about rare disorders and predisposition to common diseases at the primary care level will require the integration of different healthcare professionals into local teams. Genetic counsellors, specialist nurses and practice nurses will be needed to support the genetic practitioners.

The low priority given to genetics in the curriculum of medical schools has been widely criticised and a range of initiatives has been made to remedy this. However, because of the predicted expansion in the application of genetics to healthcare, there is an urgent need to give attention to the basic training of other healthcare professionals including nurses. How will the established general practitioner build up his or her knowledge and expertise? Involvement in the development and use of guidelines, contact with specialist centres and continuing education can play a role[54]. The adoption of pilot schemes to test out different strategies within primary and secondary healthcare will be essential.

The general public

There has also been wide discussion about the need to educate the general public about the future impact that genetics will have on their lives. In Europe, surveys have suggested that, while the majority of the public support the application of genetics and biotechnology to healthcare, they are also wary and uncertain about the wider implications. The role of the media in enhancing those uncertainties in Europe has been unhelpful. Some elements of the media have frequently been quick to play up research which aims to identify genes for intelligence, sexual preference, criminality, and even 'divorce'. This is inappropriate because genes that may be associated with these traits are unlikely to have simple phenotypic characteristic nor can social influences be eliminated.

There is a strong case to be made for initiatives that will help the public understand both the opportunities ahead for improving healthcare through the use of genetics which at the same time emphasise the likely limitations. There will always be those who seek to exploit public anxieties or weaknesses for commercial gain. However, those parents who may wish to use genetic knowledge to eliminate undesirable genes from their future offspring are likely to be in a minority. It is most important, therefore, that the public are appraised of the true limitations as well as the potential for gene identification over the next 10 years and beyond, so that exaggerated claims will not mislead or do harm to those who are vulnerable in society.

There are many initiatives which have taken place under the general heading of 'public understanding of science' to encourage the general public to think about and understand the issues associated with genetics and biotechnology as a whole[55]. These initiatives are inevitably limited in the number of people involved and their consequent impact. It will therefore be essential that a concerted effort is made through primary, secondary and tertiary education to educate children at an early stage of their lives in a responsible way. There have already been a number of calls drawing attention to this need and some initiatives are already in place. These range from information packs for schools on genetic disease to drama productions which focus on issues raised by the applcation of biotechnology. While these approaches are undoubtedly valuable they are insufficient. The wider attention to genetics in the core school curriculum is an essential goal to prepare tomorrow's adults for the future impact that genomics will certainly have on their lives.

Bibliography

1 Nuffield Council on Bioethics. *Mental Disorders and Genetics: the ethical context.* London: Nuffield Council on Bioethics, 1998

2 *Ibid*, page 3

3 *Ibid*, page 3

4 Human Genetic Advisory Commission. *The Implications of Genetic Testing for Insurance.* London: The Human Genetics Advisory Commission, 1997

5 Harper P. *Practical Genetic Counselling*, 4th edn. Oxford: Butterworth-Heinemann, 1998

6 Nuffield Council on Bioethics. *Genetic Screening: ethical issues.* London: Nuffield Council on Bioethics, 1993

7 Harper P, Clark A. *Genetics, Society and Clinical Practice.* Oxford: Bios Scientific, 1997

8 British Medical Association. *Human Genetics: choice and responsibility.* Oxford OUP, 1998

9 Nuffield Council on Bioethics. *Mental Disorders and Genetics: the ethical context.* London: Nuffield Council on Bioethics, 1998; 34

10 Michie S, Marteau T. Genetic counselling: some issues of theory and practice. In: Marteau T, Richards M. eds. *The Troubled Helix.* Cambridge: CUP, 1996

11 Craufurd D, Tyler A. Predictive testing for Huntingdon's disease: protocol of the UK Huntingdon's Prediction Consortium. *J Med Genet* 1992; **29**: 915–8

12 Richards M. Annotation: genetic research, family life, and clinical practice. *J Child Psychol Psychiatry* 1998; **39**: 291–305

13 Nuffield Council on Bioethics. *Mental Disorders and Genetics: the ethical context.* London: Nuffield Council on Bioethics, 1998; 38

14 Dudok de Wit A. *To know or not to know: the psychological implications of presymtomatic DNA testing for autosomal dominant inheritable late onset disorders.* Doctoral thesis, University of Utrecht, 1997

15 The Law Commission. *Mental Incapacity.* Law Commission No. 231. London: HMSO, 1995

16 Department of Health, Welsh Office. *Code of Practice: Mental Health Act 1983.* London: HMSO paragraph 15.8-15.24; 1993
Lord Chancellor's Department. *Who decides? Making decisions on behalf of mentally incapacitated adults.* Cm 3803, London: Lord Chancellor's Department, 1997

17 Nuffield Council on Bioethics. *Human Tissues: ethical and legal issues*. London: Nuffield Council on Bioethics, 1995

18 After the decision of the House of Lords in *Gillick v. West Norfolk and Wisbech Issues Health Area Authority* 1985 3 ALL ER 402

19 Wertz D, Fanos K, Reilly P. Genetic testing for children and adolescents: who decides? *JAMA* 1994; **272**: 878–81
Chapple A, May C, Campion P. Predictive and carrier testing of children: professional dilemmas for clinical geneticists. *Eur J Genet Sociol* 1996; **2**: 28–38
McLean S. Genetic screening of children; the UK position. *J Contemp Health Law Policy* 1995; **20**: 113–30
Michie A, Marteau T. Predictive genetic testing in children: the need for psychological research, *Br J Health Psychol* 1998; **1**: 3–14
Clarke A. ed. *Genetic Testing of Children*. Oxford: Bios Scientific, 1998

20 Nuffield Council on Bioethics. *Mental Disorders and Genetics: the ethical context*. London: Nuffield Council on Bioethics. 1998; 42

21 Clark A. *The Genetic Testing of Children: report of a working party of the Clinical Genetics Society*. London: Clinical Genetics Society, 1994

22 Green J. Steatham H. Psychological aspects of prenatal screening and diagnosis. In: Marteau T, Richards M. eds. *The Troubled Helix*. Cambridge: CUP, 1996

23 See for example: Nuffield Council on Bioethics. *Mental Disorders and Genetics: the ethical context*. London: Nuffield Council on Bioethics, 1998

24 *Ibid*, page 51

25 American College of Medical Genetics/American Society of Human Genetics Working Group on ApoE and Alzheimer disease. Statement on usefulness of apolipoprotein E testing for Alzheimer disease, *JAMA* 1995; **274**: 1627–9

26 Post S *et al* The clinical introduction of genetic testing for Alzheimer's disease, *JAMA* 1997; **277**: 832–6

27 Nuffield Council on Bioethics. *Mental Disorders and Genetics: the ethical context*. London: Nuffield Council on Bioethics, 1998

28 *Ibid*, page 25

29 Rutter M, Plomin R. Opportunities for psychiatry from genetic findings, *Br J Psychiatry* 1997; **171**: 209–19

30 William J *et al*. European Multicentre Study of Schizophrenia (EMASS) Group. *Lancet* 1996; 347: 1294–6

31 Marteau T, Croyle R. Psychological responses to genetic testing. *BMJ* 1998; **316**: 693–6

32 Rutter M, Plomin R. Opportunities for psychiatry from genetic findings. *Br J Psychiatry* 1997; **171**: 209–19

33 Nuffield Council on Bioethics. *Mental Disorders and Genetics: the ethical context*. London: Nuffield Council on Bioethics, 1998

34 Clarke A. The genetic dissection of multifactorial disease: the implications of susceptibility screening. In: Clarke A. ed. *Genetics, Society and Clinical Practice*. Oxford: Bios Scientific, 1997, 100

35 The voluntary code of practice for directing marketed tests in the UK has been drawn up by the Advisory Committee on Genetic Testing, a non-statutory Government advisory body

36 Association of British Insurers. *Genetic Testing: ABI code of practice*. London: Association of British Insurers, 1997

37 Human Genetics Advisory Commission. *The Implications of Genetic Testing for Insurance*. London: The Human Genetics Advisory Commission, 1997

38 Association of British Insurers (ABI). *Genetics: Code of Practice*. London: ABI, Press Release 17 December 1997

39 Nuffield Council on Bioethics. *Genetic Screening: ethical issues*. London: Nuffield Council on Bioethics, 1993; 64

40 House of Commons Select Committee on Science and Technology. *Human Genetics: the science and its consequences*. House of Commons 1994–95, volume 1, report and minutes of proceedings session 41-1, London: HMSO, 1995; paragraph 226

41 Holm S. Presentation to *Conference on Genetics Information: acquisition, access and control.* Preston: University of Central Lancashire, 5–6 December 1997
42 Nuffield Council on Bioethics. *Mental Disorders and Genetics: the ethical context.* London: Nuffield Council on Bioethics, 1998
43 *Ibid*, page 19
44 *Ibid*, page 5
45 Knoppers BM *et al.* Control of DNA samples and information. *Genomics* 1998; **50**: 385–401
46 Nuffield Council on Bioethics. *Mental Disorders and Genetics: the ethical context.* London: Nuffield Council on Bioethics, 1998
47 *Ibid*
48 *Ibid*, page 70.
The American Society of Human Genetics. DNA banking and DNA analysis: points to consider. *Am J Hum Genet* 1997; **8**: 781
49 *Ibid*, page 71
50 *Ibid*, page 72
51 Kinmouth AL, Reinhard J, Bobrow M. The new genetics: implications for clinical services in Britain and the United States. *BMJ* 1998; **316**: 767
52 Harper PS. Genetic testing for common diseases and health care provision. *Lancet* 1995; **346**: 1645–6
53 Nuffield Council on Bioethics. *Mental Disorders and Genetics: the ethical context.* London: Nuffield Council on Bioethics, 1998
54 Kinmouth AL, Reinhard J, Bobrow M. The new genetics: implications for clinical services in Britain and the United States. *BMJ* 1998; **316**: 767
55 Initiatives to encourage the public to familiarise themselves about genetics include the Gene Shop at Manchester Airport, citizen's juries, focus groups, consensus conferences and the work of charities such as the Genetic Interest Group in the UK

Xenotransplantation

Peter J Morris

Nuffield Department of Surgery, University of Oxford, John Radcliffe Hospital, Oxford, UK

The success of organ transplantation has led to an ever-increasing shortfall between the demand for organs and the supply. This has led to extensive investigation of the possible use of animals, especially the pig, as organ donors. However, a number of major barriers to successful xenotransplantation exist. These include immunological, physiological, anatomical, infectious and ethical problems. Of the immunological problems, the most immediate is hyperacute rejection of the organ caused by natural cytotoxic antibodies in man directed at the galactose α-1,3-galactose antigen present in all mammals except man, old world monkeys and the great apes. It is in this area in which genetic engineering has been applied most assiduously and essentially this problem is solved, at least in theory. However, many other problems remain to be resolved before xenotransplantation is likely to become a clinical reality.

The success of organ transplantation in recent years has led to an ever increasing shortfall between the demand for organs for transplantation and the supply of cadaver organs. As there seems little possibility of increasing the supply of cadaver organs, the only avenue at present available to increase the number of patients transplanted is by the increasing use of living related and unrelated donors, certainly a realistic possibility in the case of kidney transplantation, but with a limited role for other organs. In addition, if the transplantation of tissues or cells became a real possibility, *e.g.* pancreatic islets for diabetes, then again the supply of human tissues will be unable to meet the demand.

Thus this enormous discrepancy between the number of patients who require an organ transplant and the number of available organ donors has led to a major interest in the possibility of transplanting organs or tissues from animals. Indeed, the first xenograft in humans, a goat kidney into man was performed by Jaboulay (Carrel's early teacher) in 1906[1]. In the early 1960s, Reemstma transplanted kidneys from chimpanzees into patients with renal failure using azathiaprine and prednisone for immunosuppression[2]. These grafts survived from weeks to months and one suspects that with modern immunosuppression these grafts may have survived for very much longer. Other organ transplants from primate to man have been attempted from time to time but with no greater success than originally achieved in Reemstma's original

Correspondence to: Dr Peter J Morris, Nuffield Department of Surgery, University of Oxford, John Radcliffe Hospital, Headington, Oxford OX3 9DU, UK

Table 1 Some of the major barriers to xenotransplantation in man

Immunological	(i) Rejection	Hyperacute rejection Acute vascular rejection Cellular rejection Chronic rejection
	(ii) Immune response to foreign proteins	
Physiological	Incompatibility of proteins Blood flow	
Anatomical	Size	
Infectious	Bacteria Protozoa Viruses, including endogenous retroviruses	
Ethical	Selection of recipient Informed consent Use of animals	

pioneering attempts. However, for a number of reasons, including ethical reasons, which will be briefly discussed later, the use of the higher order primates (old world monkeys) is not a realistic possibility and hence any approach to clinical transplantation using animal organs must consider the use of a species that would be acceptable to a majority of patients. Such a potential donor would be the pig which is of an appropriate size and whose physiology in many instances is not too dissimilar to man. Experimental models of xenotransplantation are essential to developments in this field and such models of transplantation between species are either concordant, *i.e.* there is no pre-existing natural cytotoxic antibodies against donor species and hyperacute rejection does not occur, *e.g.* hamster heart to rat, or discordant, *i.e.* there is pre-existing natural cytotoxic antibody against the donor species and hyperacute rejection does occur, *e.g.* pig kidney to old world monkey, pig heart to man.

However, many problems have to be overcome before a discordant organ such as a pig heart could be transplanted into man. These problems are immunological, physiological, anatomical, infectious and ethical (Table 1). In this review, I will discuss some of these problems and approaches to their solution, with particular reference to the application of molecular biological techniques, concentrating on the more recent literature, as there are several comprehensive reviews of earlier work in this field[3–7]. However it must be said that most of the current activity has been directed at the definition and solution of the immunological problems, only because these will have to be largely solved before the other problems will be apparent.

Immunological problems

Hyperacute rejection

Man has natural cytotoxic antibodies to a xeno-antigen in all other species and, in particular, against the pig, the favoured donor species. This was first recognised in 1968 following the immediate failure of a pig and a sheep heart after transplantation in two patients[8]. Most of these antibodies which are both IgG and IgM were subsequently shown to be directed against a terminal carbohydrate on glycolipids or glycoproteins, galactose α-1,3-galactose(Gal), which appears to be ubiquitously expressed in most tissues in all species except man, the great apes and old world monkeys[9,10]. Hyperacute rejection of a discordant organ, such as a pig kidney in a primate (old world), is mediated by antibody, predominantly IgM, and complement. New-born human and baboon sera do not have anti-Gal IgM but only IgG, and pig heterografts in a new-born baboon do not undergo hyperacute rejection even in the presence of high titre IgG[11,12]. In contrast, transplantation of a pig organ into an adult primate (old world) does undergo hyperacute rejection[13]. Thus transplantation of a pig organ into man would result in an immediate and rapid destruction of the organ due to antibody and complement, characterised by vascular thrombosis and polymorphonuclear infiltration, just as seen in man when a transplant is carried out in a patient with cytotoxic allo-antibodies against donor histocompatibility antigens, or across an ABO incompatible blood barrier. This, then, is the first major problem to be overcome in the xenotransplantation arena and this too is the area where genetic engineering is being applied most actively (Table 2).

The initial approaches to the prevention of hyperacute rejection were directed at the removal of antibody or complement. Removal of antibody was first attempted using conventional plasmapheresis or immuno-absorption, which could reduce antibody titres and delay, but not prevent, hyperacute rejection[14]. More specific immuno-absorption either by using a pig liver connected to the circulation or by using columns coated with the gal α-1,3-gal carbohydrate were effective in substantially removing the anti-Gal antibody and preventing hyperacute rejection of a pig organ in a baboon[15–18]. However, once the immuno-absorption was completed, antibody soon returned, usually leading to delayed graft rejection. Another approach to inactivating the anti-Gal antibodies has been to infuse Gal oligosaccharides, which in a pig heart to baboon model removed antibody during infusion and, furthermore, hyperacute rejection did not occur during infusion[19]. However, as the oligosaccharides only remain in the circulation for 30 min, this approach

Table 2 Approaches to the prevention of hyperacute rejection of a xenogeneic organ, this being mediated by antibody against the Gal antigen and complement (see text for details)

Removal of antibody		
	Non-specific	Plasmapheresis or immunoabsorption
	Specific	Cross circulation with organ from donor species
		Immunoabsorption with Gal coated columns
		Infusion of Gal oligosaccharides
Removal or inhibition of complement		
		Cobra venom factor
		Soluble complement receptors
		Transgenic animals expressing human complement regulatory proteins (CD55, CD46,CD59)
Removal of Gal antigen		
		KO of galactosyl α-1,3 transferase gene
		Transgenic animals expressing α-galactosidase transgene
		Transgenic animals expressing human α-1,3 fucosyl transferase transgene
Combined technologies		
		Tg human fucosyl transferase gene + Tg α-galactosidase gene
		Tg human CD55 gene + Tg human fucosyl transferase gene
Tolerance		
		Mixed chimerism
		B cell tolerance

probably does not represent a practical approach. Nevertheless, the recent production of α-galactosyl oligosaccharides conjugated to polyethylene glycol has extended the intravascular half-life to hours and these conjugates inhibit the cytotoxic action of human sera to pig cells *in vitro* as effectively as unconjugated Gal oligosaccharides[20]. In addition, it should be noted that polymorphism of the Gal antigen is likely to make the development of an effective immunoabsorbent difficult[21].

Removal of complement can be achieved completely with cobra venom factor, which certainly leads to prevention of hyperacute rejection, but it is too toxic for clinical use. In a heterotopic pig to baboon model in which cobra venom factor was used either alone or with splenectomy and/or immunosuppression, hyperacute rejection was prevented, even though there was extensive deposition of antibody in the pig hearts, which led to failure from vascular rejection after 10 to 15 days[22]. Nevertheless it has allowed the critical role of complement in this phenomenon to be clearly defined. There is also evidence that the alternative complement pathway may be activated to cause rejection even in the virtual absence of antibody[23]. A more sophisticated approach to the inactivation of complement is the use of soluble complement receptors infused intravenously which can delay hyperacute rejection significantly, particularly if the anti-Gal antibody has been substantially reduced in titre[24,25].

The development which perhaps has brought clinical trials closest is the creation of transgenic animals expressing human complement regulatory proteins on their tissues and in particular on the endothelium, which is the primary target of the antibody and complement, mediated destruction[26]. These complement regulatory proteins are relatively species specific and, hence, the uninhibited complement-mediated destruction of porcine organs in the primate (or man) in the presence of natural cytotoxic antibodies. This has been accomplished both in the mouse and in the pig, the transgenes being CD55 (decay accelerating factor, DAF), CD46 (membrane cofactor protein, MCP), both of which inhibit complement at the C3 level, and CD59 which inhibits the late stages of complement activation at the C8–C9 level. In the mouse, one or two transgenes have been expressed with a resultant prevention of *in vitro* cell lysis by human serum of mouse target cells and delay or prevention of hyperacute rejection in a Langendorff perfusion model[27,28]. Increased expression of these regulatory proteins on endothelium has been achieved in mice by using the endothelium specific human intercellular adhesion molecule (ICAM-2) promoter in the CD55 construct which resulted in uniform expression of the human CD55 on all endothelium in the transgenic mice[29,30].

Transgenic pigs have now been produced expressing the human complement regulatory proteins where the phenomenon of hyperacute rejection of livers, kidneys, lungs and hearts has been prevented in in vitro perfusion models, and kidneys, lungs and hearts from such transgenic animals have now been transplanted into primates (cynomolgus, rhesus or baboons) without hyperacute rejection occurring[31–42]. In general, this was followed by failure from acute vascular rejection (delayed xenograft rejection), but with additional heavy immunosuppression survival for up to 60 days was achieved in some instances. However, it should be noted that, in some experiments[31], hyperacute rejection did not occur in many of the controls, suggesting that the experimental model may not always mimic what might be confidently expected if the same organ was transplanted into a human.

As the major xeno-antigen targeted by the natural cytotoxic antibodies is the Gal antigen, an obvious approach would be to produce a potential donor animal which did not express the Gal antigen. This has been achieved in the mouse by knocking out the gene for galactose α-1,3 transferase by homologous recombination. The cells from such animals are resistant to destruction by human sera *in vitro* and have not undergone hyperacute rejection in a Langendorff perfusion model[43]. However, it has so far proved impossible to produce such a knockout in the pig.

Another interesting approach, which has only been accomplished so far in the mouse, is to insert the gene for the fucosyl transferase enzyme which will then add the fucose sugar (H antigen) to surface glycoproteins or glycolipids in competition with the Gal antigen produced by the α-1,3 galactose transferase[44–46]. The H antigen is dominant in this situation, thus leading to a decreased expression of the Gal epitope. This approach can certainly suppress *in vitro* cytotoxicity of target mouse cells by human serum and would be an attractive additional approach to the prevention of hyperacute rejection in man if reproduced in the pig. Nevertheless, it should be noted that this reduction in α-Gal expression is achieved at the expense of considerable alteration of the glycosylation of the cell surface profile and with the appearance of new carbohydrate epitopes not seen in the parent cells[47].

An extension of the above was to cross mice transgenic for the human fucosyl transferase gene with mice transgenic for human α-galactosidase. The latter mice overexpress α-galactosidase which results in a functional reduction in the expression of the Gal epitope and a reduction in susceptibility to lysis by human serum. However, the combination resulted in virtually no expression of Gal on the cell surface and complete resistance to lysis by human serum in *in vitro* experiments[48].

Crossing the transgenic mice expressing the human complement regulatory protein, CD55 (DAF), with the α-galactosyl transferase (Gal) knockout mice produced mice whose cells were more resistant to complement mediated lysis than either DAF mice or Gal knockout mice alone, and whose hearts also showed longer survival when perfused *in vitro* with human serum[49,50]. This suggests that, in clinical practice, a combination of approaches is likely to be required to prevent hyperacute rejection. However, the technology is now available and it is not unreasonable to suggest that the first major barrier to xenotransplantation, namely hyperacute rejection, has now been largely overcome, at least in theory, provided some of these genetic engineering techniques can be applied successfully to the pig.

Acute vascular rejection

If hyperacute rejection can be prevented in the first 24 h, then acute vascular rejection (also known as delayed xenograft rejection) will occur. This is primarily due to endothelial activation leading to a procoagulant phase and is mediated by anti-Gal antibody. It is characterised by upregulation of adhesion molecules, platelet adhesion and neutrophil infiltration followed by thrombosis[7].

Although this may be prevented or delayed by heavy immunosuppression, this would not be acceptable in clinical practice at this time. Thus more sophisticated approaches will be required. These might include blocking of adhesion molecules expressed or upregulated on activated endothelium using antibodies to ICAM-1 or E-selectin, for example, using agents that inhibit endothelial activation or inhibiting the coagulation pathway at various levels[7]. This vascular or delayed xenograft rejection remains a major hurdle to be overcome before clinical trials can be considered.

Acute cellular rejection

Undoubtedly if both of the above can be prevented, then cellular rejection would be expected to play a role. Although it had been felt that this form of rejection might be weaker than an allograft reaction because the xeno-antigens could only be recognised by indirect antigen presentation, it is now clear that this a strong reaction mediated by T cells, NK cells and macrophages. The activation of T cells is predominantly due to indirect presentation of xenopeptides by antigen presenting cells (APC) to CD4 T cells. Mice that lack MHC class 2 on their APC will show prolonged survival of pig skin grafts which supports a major role for the CD4+ T cell[52]. The strength of this rejection reaction is illustrated by the observation that a concordant graft (where there is no anti-Gal antibody in the recipient and both donor and recipient express Gal antigen), as for example in a hamster to rat renal allograft model, is still rejected quite fiercely, usually after several days. Furthermore, this reaction is not suppressed other than modestly with conventional immunosuppressive agents such as cyclosporine[53]. This response of the rat to the hamster heart is a complex reaction for more recently it has been shown that antibodies do appear in the rat against the hamster and probably play a prominent role in the vascular rejection of the hamster heart. Furthermore, although cobra venom factor by itself did not prevent rejection in this model, the addition of cyclosporine did prevent the appearance of antibody and also prevent rejection, confirming that complement fixing antibodies do play a role in this concordant model of xenotransplantation[54]. In addition, nude rats with no T cell activity, where the B cell antibody response was blocked completely with leflunomide still rejected hamster hearts in 3 days and this was associated with a marked NK cell infiltration in the hearts[55]. But blockade of NK cells with an anti NK cell serum prolonged the survival of the hamster hearts in these Leflunomide treated nude rats.

Pig aortic endothelial cells will not stimulate human CD4 T cells to proliferate *in vitro* unless there are metabolically active human antigen presenting cells present, showing that the indirect presentation of xeno-antigen is crucial for the response to pig endothelial cells. Interestingly none of the antigen presenting cells that were adherent to cells other than endothelium were able to stimulate CD4 T cells to proliferate despite presumably providing a full set of xenogeneic peptides, suggesting that the pig endothelium must provide some specific stimulatory signal for this proliferative response[56].

Other relevant experiments were carried out in a new-born baboon model given pig hearts. The new-born baboons do not hyperacutely reject pig hearts as they do not have anti-Gal IgM antibodies. Nevertheless, the hearts are rejected in 3–4 days, and this is associated with a marked infiltration of macrophages and NK cells. Use of conventional immunosuppressive drugs extended survival by a few days and it was noted that anti-Gal IgM appeared[57]. Thus it seems probable in clinical practice that quite powerful immunosuppression will be required to suppress this part of the xenoreactive response.

Chronic rejection

Little is known about this form of rejection in the xenograft model, but it seems likely that if the early forms of rejection are suppressed, then chronic rejection changes are inevitable and these will not be prevented by current immunosuppression, especially if they result from reactions produced by the anti-Gal antibody. These changes will resemble those seen in chronic allograft rejection which are characterised by myo-intimal hypertrophy of arteries and interstitial fibrosis. However, this remains an unknown quantity in xenotransplantation at this time.

Tolerance

The achievement of tolerance to a xenogeneic organ would be the ultimate goal in xenotransplantation just as it is in the allogeneic arena. Indeed, successful clinical xenotransplantation may never be achieved until tolerance to a xenogeneic organ can be produced. Already an approach leading to B cell tolerance to Gal antigens in the rodent has been described[51]. Major advances have been made by Sachs' group in recent years in a pig to primate model where mixed chimerism has been produced after an ablative preparation of the recipient has been carried out. This has resulted in prolonged survival of a subsequently transplanted kidney[58]. Similarly, in the mouse, lasting mixed chimerism has been produced in Gal knockout mice by transplanting a mixture of Gal knockout and wild type bone marrow after lethal irradiation of the knockout mice. After several weeks, no anti-Gal antibody was detected and this was a long lasting effect[59].

A somewhat different approach recently described involved the induction of B cell tolerance to the Gal antigen. This involved the introduction of a functional porcine α-glycosyl transferase gene into murine bone marrow cells using a retroviral vector. These transduced bone marrow cells were used to reconstitute Gal knockout mice (which do not express Gal antigen and hence make antibody against Gal) after lethal irradiation. The reconstituted mice did not produce anti-Gal antibodies suggesting that tolerance at a B cell level had been produced[51]. Thus, this approach, involving the production of molecular chimerism in autologous bone marrow, might be an attractive approach to the induction of tolerance to the Gal antigen in clinical practice.

Physiological

Physiological problems essentially remain unknown at this time, but some idea that these may be considerable can be gleaned from the baboon liver to man transplants and the pig kidney to primate transplants[60,31]. In the former, for example, the cholesterol level in the patient remained extremely low, at the baboon level rather than approaching the lower limits of human levels. In the latter, the primates with a transgenic pig kidney became profoundly anaemic within 60 days presumably because pig erythropoietin is not compatible with receptors on the human erythroid stem cells. Undoubtedly, many foreign proteins produced by surviving xenotransplants will be incompatible with their human ligands. Furthermore, there is very likely to be an immune response to any foreign protein produced by the xenogeneic organ, and this is an equally plausible explanation for the anaemia seen in the primates with a pig kidney.

Anatomical

Size will obviously be important in choosing organs for human use but, in this respect, the pig should be reasonably satisfactory in most instances[61]. However, will some organs from the quadruped pig, for example, be compromised by failure to adjust blood flows appropriately in upright man? In general, the anatomical structure of the pig is not very different to that of man.

Infectious

This is potentially a serious problem for, although pathogen-free pigs could be bred so as to be free of bacteria and protozoa, the recent description of endogenous porcine retroviruses and the demonstration that these retroviruses could be transferred *in vitro* to human cells has given rise to considerable anxiety[62,63]. Although these endogenous retroviruses are not associated with disease in the pig, their transfer to a heavily immunosuppressed patient might give rise to a new disease entity and, furthermore, this might lead to the introduction of a new pathogen into the population. These initial observations have led to the careful examination of patients who have been exposed to any pig tissues, *e.g.* pig pancreatic islets and, to date, no evidence for the transfer of a porcine retrovirus has been found, but of course the number of patients studied remains relatively small[64,65]. There is also a report of the transplantation of pig aortic endothelial cells into heavily immunosuppressed baboons in whom 24 months later no evidence of transmission of virus could be demonstrated[66]. On the other hand, a novel swine hepatitis E virus has been described which also might have implications for the use of pigs as donors[67]. In addition, retrospective analysis of tissues from two patients who received a baboon liver transplant and survived 27 and 70 days, respectively, revealed the presence of two simian retroviruses of baboon origin[68]. Thus the possible transmission of viruses from xenogeneic donors to immunosuppressed recipients remains of considerable concern in future clinical trials of xenotransplantation.

Ethical

Although organs from the great apes would probably behave in a similar manner to human allografts with modern immunosuppression, at least in the short term, it seems that their use would not be generally approved for ethical reasons. But, even if this were not the case, the numbers that could be bred in captivity would be very limited, they are an endangered species and the risk of transmitting viral infections would be considerable. Although some people would find the use of any animals ethically unacceptable, most would probably accept the use of a species that was bred as a food source *e.g.* pig, sheep and cattle, although perhaps finding a certain degree of repugnance in receiving an organ from such a source. What is more likely to provide significant ethical problems if xenotransplantation becomes a clinical reality is the choice that inevitably would have to be made between a xenograft and an allograft for a given patient. This would appear to limit the early trials of organ xenotransplantation to patients who are not eligible for an allograft or have a truly emergent need for an organ transplant with no available human allograft. Furthermore, can informed consent be truly given in this situation, given the uncertainty which will surround the early trials and the obligation the patient will have to undergo continuing and long-term surveillance for evidence of viral infection transmitted from the donor animal[69].

Key points for clinical practice

Although no-one can doubt the need for successful xenotransplantation of organs and tissues in the management of end-stage organ or tissue failure, there should be no illusions about the obstacles to be overcome before this can happen, as these indeed remain very considerable. Perhaps, in the future, the successful cloning of human tissues and organs might become possible, in which case xenotransplantation would no longer be needed.

References

1 Hamilton D. Kidney transplantation: a history. In: Morris PJ, ed. *Kidney Transplantation: Principles and Practice*, 4th edn. Philadelphia: Saunders, 1994: 1–7

2 Reemstma K, McCracken BH, Schlegal JU *et al.* Renal heterotransplantation in man. *Ann Surg* 1964; **160**: 384-XX

3 Auchincloss H. Xenogeneic transplantation. A review. *Transplantation* 1988; **46**: 1–20

4 Auchincloss H. Xenografting: a review. *Transplant Rev* 1990; **4**: 14–27

5 Bach FH, Dalmasso G, Platt JL. Xenotransplantation: a current perspective. *Transplant Rev* 1992; **6**: 163–75

6 Bach FH, Auchincloss H, Robson SC. Xenotransplantation. In: Bach FH, Auchincloss H, eds. *Transplantation Immunology*. New York: Wiley-Liss, 1995; 305–38

7 Bach FH, Ferran C, Soares M *et al.* Modification of vascular responses in xenotransplantation: inflammation and apoptosis. *Nat Med* 1997; **3**: 944–8

8 McKenzie IFC, Stocker J, Ting A, Morris PJ. Human lympho-cytotoxic and haemagglutinating activity against sheep and pig cells. *Lancet* 1968; **ii**: 386–7

9 Galili U, Shohet SB, Kobrin E, Stults CLM, Macher BA. Man, apes and old world monkeys differ from other mammals in the expression of α-galactosyl epitopes on nucleated cells. *J Biol Chem* 1988; **263**: 17755–62

10 Sandrin M, Vaughan HA, McKenzie IFC. Identification of Gal(α1,3)Gal as the major epitope for pig-to-human vascularised xenografts. *Transplant Rev* 1994; **8**: 134–49

11 Minanov OP, Itescu S, Neethling FA *et al.* Anti-Gal IgG antibodies in sera of newborn humans and baboons and its significance in pig xenotransplantation. *Transplantation* 1997; **63**: 182–6

12 Minanov OP, Artrip JH, Szabolcs M *et al.* Triple immunosuppression reduces mononuclear cell infiltration and prolongs graft life in pig-to-newborn baboon cardiac xenotransplantation. *J Thorac Cardiovasc Surg* 1998; **115**: 998–1006

13 Lambrigts D, Sachs DH, Cooper DK Discordant organ xenotransplantation in primates: world experience and current status. *Transplantation* 1998; **66**: 547–61

14 Pascher A, Poehlein C, Stangl M *et al.* Application of immunoapheresis for delaying hyperacute rejection during isolated xenogeneic pig liver perfusion. *Transplantation* 1997; **63**: 867–75

15 Taniguchi S, Neethling FA, Korchagina EY et al. *In vivo* immunoadsorption of antipig antibodies in baboons using a specific Gal(alpha)1-3Gal column. *Transplantation* 1996; **62**: 1379–84

16 Kozlowski T, Lerino FL, Lambrigts D *et al.* Depletion of anti-Gal(alpha)1-3Gal antibody in baboons by specific alpha-Gal immunoaffinity columns. *Xenotransplantation* 1998; **5**: 122–31
17 Lin SS, Kooyman DL, Daniels LJ *et al.* The role of antural anti-Gal alpha 1-3Gal antibodies in hyperacute rejection of pig-to-baboon cardiac xenotransplants. *Transpl Immunol* 1997; **5**: 212–8
18 Xu Y, Lorf T, Sablinski T *et al.* Removal of anti-porcine antural antibodies from human and nonhuman primate plasma *in vitro* and in vivo by a Gal-alpha-1-3-Gal-beta-1-4-beta-Glc-X immunoaffinity column. *Transplantation* 1998; **65**: 172–9
19 Simon PM, Neethling FA, Taniguchi S *et al.* Intravenous infusion of Gal-alpha 1-3Gal oligosaccharides in baboons delays hyperacute rejection of porcine heart xenografts. *Transplantation* 1998; **65**: 346–53
20 Nagasaka T, Kobayashi T, Muramatsu H *et al.* Alpha-galactosyl oligosaccharides conjugated with polyethylene glycol as potential inhibitors of hyperacute rejection upon xenotransplantation. *Biochem Biophys Res Commun* 1997; **232**: 731–6
21 McKane W, Lee J, Preston R *et al.* Polymorphism in the human anti-pig natural antibody repertoire: implications for antigen-specific immunoadsorption. *Transplantation* 1998; **66**: 626–33
22 Kobayashi T, Taniguchi S, Neethling FA *et al.* Delayed xenograft rejection of pig-to-baboon cardiac transplants after cobra venom factor therapy. *Transplantation* 1997; **64**: 1255–61
23 Romanella M, Aminian A, Adam WR, Pearse MJ, d'Apice AJ. Involvement of both the classical and alternate pathways of complement in an *ex vivo* model of xenograft rejection. *Transplantation* 1997; **63**: 1021–5
24 Pruitt SK, Kirk AD, Bollinger RR *et al.* The effect of soluble complement receptor type 1 on hyperacute rejection of porcine xenografts. *Transplantation* 1994; **57**: 363–XX
25 Pruitt SK, Bollinger RR, Collins BH *et al.* Effect of continuous complement inhibition using soluble complement receptor Type I on survival of pig-to-primate xenografts. *Transplantation* 1997; **63**: 900–2
26 Platt JL, Logan JS. Use of transgenic animals in xenotransplantation. *Transplant Rev* 1996; **10**: 69–77
27 Van Denderen BJ, Pearse MJ, Katerelos M *et al.* Expression of functional decay-accelerating factor (CD55) in transgenic mice protects against human complement-mediated attack. *Transplantation* 1996, **61**: 582–8
28 Yannoutsos N, Ijzermans JN, Harkes C *et al.* A membrane cofactor protein transgenic mouse model for the study of discordant xenograft rejection. *Genes Cells* 1996; **1** 409–19
29 Cowan PJ, Shinkel TA, Witort EJ, Barlow H, Pearse MJ, d'Apice AJ. Targeting gene expression to endothelial cells in transgenic mice using the human intercellular adhesion molecule 2 promoter. *Transplantation* 1996; **62**: 155–60
30 Cowan PJ, Somerville CA, Shinkel TA *et al.* High-level endothelial expression of human CD59 prolongs heart function in an *ex vivo* model of xenograft rejection. *Transplantation* 1998; **65**: 826–31
31 Zaidi-A, Schmoeckel M, Bhatti F *et al.* Life-supporting pig-to-primate renal xenotransplantation using genetically modified donors. *Transplantation* 1998; **65**: 1584–90
32 Pascher A, Poehlein C, Storck M *et al.* Immunopathological observations after xenogeneic liver perfusions using donor pigs transgenic for human decay-accelerating factor. *Transplantation* 1997; **64**: 384–91
33 Schmoeckel M, Nollert G, Shahmohammadi M *et al.* Transgenic human decay accelerating factor makes normal pigs function as a concordant species. *J Heart Lung Transplant* 1997; **16**: 758–64
34 Daggett CW, Yeatman M, Lodge AJ *et al.* Swine lungs expressing human complement-regulatory proteins are protected against acute pulmonary dysfunction in a human plasma perfusion model. *J Thorac Cardiovasc Surg.* 1997; **113**: 390–8
35 Storck M, Abendroth D, Prestel R *et al.* Morphology of hDAF (CD55) transgenic pig kidneys following *ex-vivo* hemoperfusion with human blood. *Transplantation* 1997; **63**: 304–10
36 Byrne GW, McCurry KR, Martin MJ, McClellan SM, Platt JL, Logan JS. Transgenic pigs expressing human CD59 and decay-accelerating factor produce an intrinsic barrier to complement-mediated damage. *Transplantation* 1997; **63**: 149–55

37 Schmoeckel M, Nollert G, Shahmohammadi M *et al.* Prevention of hyperacute rejection by human decay accelerating factor in xenogeneic perfused working hearts. *Transplantation* 1996; **62**: 729–34

38 Diamond LE, McCurry KR, Martin MJ *et al.* Characterization of transgenic pigs expressing functionally active human CD59 on cardiac endothelium. *Transplantation* 1996; **61**: 1241–9

39 Yeatman M, Daggett CW, Parker W *et al.* Complement-mediated pulmonary xenograft injury: studies in swine-to-primate orthotopic single lung transplant models. *Transplantation* 1998; **65**: 1084–93

40 Lin SS, Weidner BC, Byrne GW et al. The role of antibodies in acute vascular rejection of pig-to-baboon cardiac transplants. *J Clin Invest* 1998; **101**: 1745–56

41 Daggett CW, Yeatman M, Lodge AJ *et al.* Total respiratory support from swine lungs in primate recipients. *J Thorac Cardiovasc Surg* 1998; **115**: 19–27

42 Schmoeckel M, Nollert G, Shahmohammadi M *et al.* Transgenic human decay accelerating factor makes normal pigs function as a concordant species. *J Heart Lung Transplant* 1997; **16**: 758–64

43 Tearle RG, Tange MJ, Zannettino ZL *et al.* The alpha-1,3-galactosyltransferase knockout mouse. Implications for xenotransplantation. *Transplantation* 1996; **61**: 13–9

44 Cohney S, McKenzie IF, Patton K *et al.* Down-regulation of Gal alpha(1,3)Gal expression by alpha1,2-fucosyltransferase: further characterization of alpha1,2-fucosyltransferase transgenic mice. *Transplantation* 1997; **64**: 495–500

45 Shinkel TA, Chen CG, Salvaris E *et al.* Changes in cell surface glycosylation in alpha 1,3-galactosyltransferase knockout and alpha1,2-fucosyltransferase transgenic mice. *Transplantation* 1997; **64**: 197–204

46 Chen CG, Salvaris EJ, Tomanella M *et al.* Transgenic expression of human alpha1,2-fucosyltransferase (H-transferase) prolongs mouse heart survival in an ex vivo model of xenograft rejection. *Transplantation* 1998; **65**: 832–7

47 Sepp A, Skacel P, Lindstedt R, Lechler RI. Expression of alpha-1,3-galactose and other type 2 oligosaccharide structures in a porcine endothelial cell line transfected with human alpha-1,2-fucosyltransferase cDNA. *J Biol Chem* 1997; **272**: 23104–10

48 Osman N, McKenzie IF, Ostenried K, Ionnou YA, Desnick RJ, Sandrin MS. Combined transgenic expression of alpha-galactosidase and alpha1,2-fucosyltransferase leads to optimal reduction in the major xenoepitope Gal alpha(1,3)Gal. *Proc Natl Acad Sci USA* 1997; **94**: 14677–82

49 Van Denderen BJ, Salvaris E, Tomanella M *et al.* Combination of decay-accelerating factor expression and alpha1,3-galactosyltransferase knockout affords added protection from human complement-mediated injury. *Transplantation* 1997; **64**: 882–8

50 Cowan PJ, Chen CG, Shinkel TA *et al.* Knock out of alpha 1,3-galactosyltransferase or expression of alpha1,2-fucosyltransferase further protects CD55- and CD59-expressing mouse hearts in an ex vivo model of xenograft rejection. *Transplantation* 1998; **65**: 1599–604

51 Bracy JL, Sachs DH, Lacomini J. Inhibition of xenoreactive natural antibody production by retroviral gene therapy. *Science* 1998; **281**: 1845–7

52 Chitilian HV, Laufer TM, Stenger K, Shea S, Auchincloss Jr H. The strength of cell-mediated xenograft rejection in the mouse is due to the CD4+ indirect response. *Xenotransplantation* 1998; **5**: 93–8

53 Homan WP, Williams KA, Fabre JW, Millard PR, Morris PJ. Prolongation of cardiac xenograft survival in rats receiving cyclosporin A. *Transplantation* 1981; **31**: 164–6

54 Miyatake T, Sato K, Takigami K *et al.* Complement-fixing elicited antibodies are a major component in the pathogenesis of xenograft rejection. *J Immunol* 1998; **160**: 4114–23

55 Lin Y, Vandeputte M, Waer M. Natural killer cell- and macrophage-mediated rejection of concordant xenografts in the absence of T and B cell responses. *J Immunol* 1997; **158**: 5658–67

56 Vallee I, Guillaumin JM, Thibault G *et al.* Human T lymphocyte proliferative response to resting porcine endothelial cells results from an HLA-restricted, IL-10-sensitive, indirect presentation pathway but also depends on endothelial specific co-stimulatory factors. *J Immunol* 1998; **161**: 1652–8

57 Minanov OP, Artrip JH, Szabolcs M *et al.* Triple immunosuppression reduces mononuclear cell infiltration and prolongs graft life in pig-to-newborn baboon cardiac xenotransplantation. *J Thorac Cardiovasc Surg* 1998; **115**: 998–1006

58 Sablinski T, Gianello PR, Bailin M *et al.* Pig-to-monkey bone marrow and kidney xenotransplantation. *Surgery* 1997; **121**: 381–91
59 Yang YG, deGoma E, Ohdan H *et al.* Tolerization of anti-Gal alpha 1-3 Gal natural antibody-forming B cells by induction of mixed chimerism. *J Exp Med* 1998; **187**: 1335–42
60 Starzl TE, Fung J, Tzakis A *et al.* Baboons to human liver transplant. *Lancet* 1991; **341**: 65–71
61 Hammer C, Linke R, Wagner F, Diefenbeck M. Organs from animals for man. *Int Arch Allergy Immunol* 1998; **116**: 5–21
62 Patience C, Takenchi Y, Weiss RA Infection of human cells by an endogenous retrovirus of pigs. *Nat Med* 1997; **3**: 282–6
63 Martin U, Kiessig V, Blusch JH *et al.* Expression of pig endogenous retrovirus by primary porcine endothelial cells and infection of human cells. *Lancet* 1998; **352**: 692–4
64 Patience C, Patton GS, Takeuchi Y *et al.* No evidence of pig DNA or retroviral infection in patients with short-term extracorporeal connection to pig kidneys. *Lancet* 1998; **352**: 699–701
65 Heneine W, Tibell A, Switzer WM *et al.* No evidence of infection with porcine endogenous retrovirus in recipients of porcine islet-cell xenografts. *Lancet* 1998; **352**: 695–9
66 Martin U, Steinhoff G, Kiessig V *et al.* Porcine endogenous retrovirus (PERV) was not transmitted from transplanted porcine endothelial cells to baboons *in vivo*. *Transpl Int* 1998; **11**: 247–51
67 Meng XJ, Purcell RH, Halbur PG *et al.* A novel virus in swine is closely related to the human hepatitis E virus. *Proc Natl Acad Sci USA* 1997; **93**: 9860–5
68 Allan JS, Broussard SR, Michaels MG *et al.* Amplification of simian retroviral sequences from human recipients of baboon liver transplants. *AIDS Res Hum Retroviruses* 1998; **14**: 821–4
69 Daar AS. Ethics of xenotransplantation: animal issues, consent, and likely transformation of transplant ethics. *World J Surg* 1997; **21**: 975–82

The impact of genetics on medical education and training

Robin Fears*, David Weatherall[†] and **George Poste***

**SmithKline Beecham Pharmaceuticals, Essex, UK*
[†]*Institute of Molecular Medicine, University of Oxford, Oxford, UK*

This paper explores, mainly from the UK perspective, some of the issues relating to the current, and potential, impact of advances in genetics and molecular biology on the education and research training of healthcare professionals. We start by describing some of the expectations for progress in the use of genomic technologies and genetic data in healthcare delivery and the need for policy development to ensure timely translation of advances in science and technology into improved patient care. We review briefly the likely evolution of clinical genetics service provision to build the requisite scientific basis in primary care and explore how user needs could be addressed. Strategic issues for the future medical curriculum are introduced and linked with the concerns about the current status of clinical academic research. The issues for research training, career progression, nurturing of research 'at the bedside', definition of the research agenda and weaknesses in both academic infrastructure and support costs are reviewed in the context of the urgent imperative for medicine to harness the accelerating pace of progress in genomics.

Genomics and molecular medicine

Genomics, the study of the genetic control of body function in health and disease, has profound implications for biomedical R&D and for the future delivery of healthcare services. One expected development will be an improved understanding of the fundamental mechanisms underlying disease, with accompanying gain in the diagnosis, prevention and therapy of major diseases. The growth in genomics also promises to redefine medicine in other ways, most notably in the areas of diagnostic testing for improved disease detection, disease staging and patient stratification, plus eventual evolution of the means to identify predisposition to disease and more precise subtyping of heterogeneous diseases[1].

Correspondence to: Dr Robin Fears, SmithKline Beecham Pharmaceuticals, New Frontiers Science Park, Harlow, Essex CM19 5AW, UK

Molecular genetics was applied initially in medicine to characterise the monogenic Mendelian disorders, but efforts have turned to the elucidation of the genetic basis of the common diseases of multigenic origin[1]. By determining the genetic component of intractable disorders like coronary artery disease, stroke, diabetes, rheumatism and cancer, it should be possible to find out much more precisely why they are caused and the basis for the various environmental interactions which are involved. This, in turn, should revolutionise the pharmacological approach to their management and, in addition, allow public health measures to be focused more effectively on particularly susceptible subsets of the community. In addition, work on somatic cell genetics, that is genetic changes acquired during our lifetimes, promises to revolutionise the diagnosis and management of many common cancers and, may also provide invaluable information about the mechanisms of ageing. Examples of the potential for the new genetics and the underpinning role of technology development are described in detail in the *Reports of the Genetics Research Advisory Group*[2,3] to the NHS Central Research and Development Committee and elsewhere[4,5]. Genetics R&D is expected to make a major contribution in addressing many of the health targets (heart disease, stroke, cancer, mental health) identified in the recent Green Paper *Our Healthier Nation*[6].

The increased knowledge of the basic biology of disease will eventually transform medical practice[7]. There will be a significant transition from our current reliance on the diagnosis and treatment of overt disease to a greater emphasis on the prediction and prevention of covert disease[8]. Consideration of the role of genes in defining susceptibility to disease is, however, too narrow a perspective – and parallel progress will be needed to understand the complex interplay of gene loci that determine disease progression, disease complications and response to treatment, plus how the penetrance of risk alleles can be increased or attenuated by environmental factors.

Emerging implications for clinical practice and training

These advances in molecular biology will undoubtedly affect the delivery of healthcare provision, at a time when such matters are already under considerable public scrutiny[9]. While the introduction of genetic information will have profound benefits for the health service, good clinical practice must not be allowed to suffer by adoption of an excessively mechanistic approach that ignores the clinical and social benefits of an holistic approach and provision of pastoral care[10]. Significant changes will be necessary in the framework for healthcare delivery, the medical curriculum and clinical research training. These elements of service provision and development are already under intense pressure. As noted by the House of Commons Science and Technology Committee Inquiry, the potential impact of R&D in genetics has been underestimated at the policy level[11]. In short, the accelerating momentum of genetics in medicine will have dramatic implications for all healthcare professionals.

The impact of genetics R&D on healthcare delivery and training , and on the role of healthcare professionals in educating the public at large on the benefits of disease prediction and prevention, provides a major challenge for those charged with guiding the introduction of new and emerging medical technologies. A delicate balance must be forged between the legitimate need to control technology diffusion while research is conducted to assess clinical and cost effectiveness versus pressures for adoption created by the media, public demand, producers and the profession[12]. The continuing role, and relevance, of UK national initiatives such as the Standing Group on Health Technologies, National Co-ordinating Centre for Health Technology Assessment, National Screening Committee and specific review bodies such as the Gene Therapy Advisory Committee, Advisory Committee on Genetic Testing, Human Genetics Advisory Commission need to be assessed on a continuing basis in the context of providing an integrated and coherent framework for the introduction of genetic-based products and services, and the training of healthcare professionals.

Building clinical genetics service provision

The organisation and manpower needs of clinical genetics services are described in the Report of the Royal College of Physicians[13]. It is beyond the scope of the present paper to examine the detail of planning human resources. As the increasing application of genetics to common disorders emerges in service delivery, the role of the consultant clinical geneticist will remain important for diagnosis and counselling. However, there is an urgent imperative to ensure appropriate incorporation of conceptual and practical elements of genetics and molecular medicine in undergraduate and postgraduate education. Links with primary care[3,11] and public health medicine will be fundamental in achieving full and equitable delivery of genetic services. To date, UK genetics services have been organised primarily at the local/regional level with a strong R&D interaction. However, as testing develops for the predisposition to multifactorial disorders, the volume of activity will grow such that different models of service delivery must be explored[2,7]. Models that allow both NHS and commercial development of large-scale testing with adequate quality control, legal oversight and clinical input, so as not to divorce testing from pre-and post-test counselling, need to be examined[7]. Clarification of the primary care team approach will be of central importance in defining the demands for professional education in genetics[14,15].

The goals of medical genetics can also only be fulfilled in the context of an educated, informed public[15]. This objective has major implications for the secondary school curriculum and for the development of ways to build public confidence in, and understanding of, the new genetics[14,16]. In the absence of this informed dialogue and educational underpinning, there is likely to be both unrealistic optimism and unrealistic fear about the new technologies.

A recent comparison of the contrasting developments in clinical genetics services in the UK and US[14] suggests that each offers special opportunities in innovation and evaluation. While a range of pilot schemes have been set up in the UK to explore how genetics services can be incorporated into mainstream clinical medicine, few have yet been fully evaluated.

Articulating user wants

In their analysis of the organisation of UK services, the Genetic Interest Group[17] concluded that medical genetics was distinct from other rapidly-advancing fields in that the unit of concern was the family and not just the individual; there was a greater emphasis on preventative medicine; and outcomes are difficult to measure in traditional ways. This last feature will remain a problem – what constitutes an index of clinical effectiveness in genetics? In the absence of clear understanding of what is meant by effectiveness (and the embodiment of this understanding in education programmes), it will continue to be difficult to resolve the major problems in UK genetic services currently perceived by users[18]:

- Access is not always consistent, or satisfactory.
- Genetic services are not always adequately linked into other medical specialities.
- There is a shortage of resource to bring new genetics-based techniques into mainstream health service provision.

Patient interest groups covering the rare genetic disorders emphasise that services should be commissioned on a regional basis, but, as genetic information becomes relevant to the common disorders, user wants will change and the healthcare educational implications will broaden. A Citizens' Jury was recently organised to begin to define the user's agenda of issues in genetic testing for susceptibility to common diseases. Among the training recommendations[18] were:

- Planning for the future should be much higher on the agenda of policy-makers and professionals at all levels within the NHS.
- Genetics should have a more prominent part in the national curriculum and community education.
- Adequate resources should be allocated to appropriate training of all involved NHS staff.
- Basic training in genetics for NHS staff should meet nationally-agreed standards – perhaps a uniform curriculum at prequalification stage.

Developing the medical curriculum

When considering the impact of genetics and molecular biology on the curriculum, it will be necessary to fulfil a number of challenging objectives[4]: continuing education for serving professionals; the education of new professionals; information dissemination to professionals; and the R&D agenda.

In particular, in addition to considering the needs of clinical genetics as a speciality, it is also necessary to consider how the broader teaching of genetics will be organised and how it will become an integral part of all the medical specialities. We recommend the following changes:

- Inclusion of genetic principles throughout the medical school curriculum rather than treating as a separate subject in the preclinical or clinical years.
- Suitable courses to be made available as part of postgraduate training programmes for clinicians and other health workers and as a duty of all major teaching centres.
- Some component of genetics training should be included in the continuing education programmes for doctors of all specialities.

In the US, where consideration of some of these issues is more advanced, a variety of initiatives is underway. While genetics is currently taught at all US medical schools, a minimum competency is not stipulated and it is recognised that there is a serious, and growing, gap in the ability of GP family physicians to both be aware of new advances and to provide relevant consultation[19]. There is a new sense of urgency to bring genetic literacy to all practising physicians (and other health professionals)[20,21]. In addition to initiatives to develop the medical school curriculum so that genetics education is fully integrated, there are many related proposals to develop Web-based resources and CD-ROM formats for continuing education programmes[19,20].

The brevity of this paper precludes a review of the other initiatives in many countries or detailed considerations of desired content, but future evolution of the medical curriculum in genetics and molecular medicine must emphasise the following:

- The importance of the new interdisciplinary framework to integrate molecular biology and clinical genetics with pathology, epidemiology and computing; the convergence of genomics and informatics in particular heralds a new era of biomedical research[22].

- The changing focus in the role of genetics not just in studying rare Mendelian diseases but encompassing the genetics of common disorders, in terms of the correlation between specific genes and disease predisposition, the definition of the molecular basis of heterogeneity in disease severity and progression, and understanding of how genetic variability can alter responsiveness to drugs and environmental factors.

- The principles and methods of genetic counselling, to address the complex range of ethical, legal and social issues posed by genetics, particularly screening of health asymptomatic individuals for predisposition to future disease[23].

- The bio-ethical dimension of genomics has evoked substantial debate[15,16]. Concern about such issues as genetic discrimination, neurogenetic determinism, eugenics and germ line gene therapy and cloning have crystallised concerns within various constituencies about the pace and scale of technological change[8]. Following *Tomorrow's Doctors*, medical ethics and law have become a core component of the medical curriculum and a recent consensus statement from the Group of Teachers of Medical Ethics and Law in UK Medical Schools[24] provides a central place for the new genetics and for related issues appertaining to confidentiality, informed consent, reproduction, medical research, resource allocation, *etc.*

The priority, however, is not just to change the course content, albeit this is vital. What is also needed is to prepare doctors for the changing culture of the clinical transaction arising, for example, in the use of disease predisposition data by hitherto healthy individuals[25].

Status of clinical research

Identifying the problems

There is no doubt that the scale and pace of advances in genetics R&D, and the financial and organisational implications for service delivery, will have considerable impact at all levels of medical education and research training. These increasing demands for translating advances in basic biomedical research into advances in patient care come at a time when there are substantial constraints on the resources allocated for clinical research and training. These general issues for the science-medicine interface were identified in the UK Task Force Report on Clinical Academic Careers[26] and reviewed in the House of Lords Science and Technology Committee Inquiry[27]. The recent increase in number of medical students proposed by the Medical Workforce Standing Advisory Committee[28], if funded on a marginal cost basis, will considerably increase the teaching and training burden, providing yet further disincentives to careers in clinical research.

Thus, even if the curriculum for medical education in genetics develops appropriately, will we lose the capability to train clinical researchers? These worries for genetics clinical research training are superimposed on various concerns raised by the Research Councils in their recent input to the UK Government's Comprehensive Spending Review. For example, there is current weakness in the integration of basic medical, clinical and social sciences research. There has been a failure to develop bioinformatics and other costly advanced computational tools and infrastructure, and there is a continuing need to establish clear guidelines to support the transparent and ethical uses of genetic databases for clinical decisions.

In a recent discussion convened[29] to address some of these issues, research and training concerns raised by the clinical academic community were reinforced by evidence from the other healthcare R&D partners. Patient interest groups, while recognising the substantial declared commitment of the NHS to clinical research, expressed dismay at an ethos of health service conformity, the apparent lack of innovation, and lack of strategy to control the variability in supply of novel research-based products and services. Medical research charities were suffering in consequence of the poor public perception of clinical research, and the pharmaceutical industry felt increasingly deterred from investing in UK clinical research. Further detail on these perspectives and on the following points can be found in the Report[29].

Nurturing clinical research

There are several issues of concern relating to sustaining a cadre of clinical researchers of international rank, how to integrate basic and clinical research, and how best to provide the training and incentives to create excellent teachers. Basic research must urgently address the needs of the post-genomic era by building interdisciplinary frameworks to link molecular biology to physiology, pharmacology and pathology, as well as develop relevant algorithms for population genetics and the analysis of geno-phenotypic relationships in the major human diseases. But, the greatest problems in biomedical research currently are 'at the bedside'. The challenge is to create an environment to support patient-oriented research at a time of increasing workload pressures arising from service delivery, increased educational demands and resource constraints. These issues are global and analogous problems are emerging in the US with the advent of Managed Care and the accompanying financial and performance demands being placed on academic medical centres.

Defining the research agenda

The initial creation of the NHS R&D function was a significant and praiseworthy event. We now need to do even better and other voices should be heard in setting the agenda. New concordats might be developed to include patients, research charities and industry, and public awareness of the benefits of clinical research must be promoted.

The most effective mechanisms for organising research for those funded from the higher education budget remains an important problem. Should every medical school cover all research areas or should some specialise and others do nothing? If there is to be a coherent promotion of selectivity in research, how should the national policy be developed?

Within the overall research agenda, there are several important policy and funding issues to address as a consequence of changing styles of research. New sources of data in genetics health research are dramatically increasing in value – for example, population databases and archived issue banks. These developments raise issues of relevance to the developing curriculum on ethics and social aspects: relating to informed consent, access, confidentiality of data and intellectual property rights.

Research infrastructure and support costs

There are considerable strains in the UK dual funding system, as confirmed by the National Committee of Inquiry on Higher Education[30]. Furthermore, the funding environment for bedside research is deteriorating as a consequence of the emphasis on laboratory research resulting from the last Research Assessment Exercise (RAE), and the relatively poor performance of clinical departments in the RAE. There is an urgent case to be made for increased government funding but also to find better ways of using current resources for research and training within both the Funding Council and NHS funding streams. For example, in primary care, much of the funding supporting professional development for individuals might be better directed to practice development for primary care teams[31].

In an era characterised by the need for major investment to drive state-of-the-art biomedical research, it is also important to establish how best to share resources among teaching hospitals. Centralisation in multidisciplinary research centres[10] provides effective coverage but must be progressed with care to avoid conflict with teaching priorities.

Training medical students and research career structure

The Task Force on Clinical Academic Careers[26] emphasised the importance of exploring new options in research training, in particular the intercalated BSc and the combined medical PhD pathway. The decrease in funding for the intercalated BSc is worrying, although some question its value[29], being based, inevitably, on a didactic rather than research emphasis. The newer joint medical PhD proposal could be valuable in encouraging substantial research experience but even the well-regarded US MD-PhD programmes suffer the potential weakness of being overly laboratory-based and not inspiring the return to clinical medicine. The effective way of training most clinical researchers is to allow 3 years of adequate fellowship training following qualification and basic training; other mechanisms to provide formal postgraduate training in genomics (for example, a MSc course) for clinicians in all specialities and general practitioners should also be considered.

In addition, in spite of Department of Health efforts (for example, *A Guide to Specialist Registrar Training*[32]), there is still uncertainty within the medical profession about the flexibility of the entry/exit points in academic career progression when embarking on professional qualifications for the royal medical colleges. This rigidity also creates problems, for example, for industry, in promoting transfer between academic and other career options. There are also problems of poor research training and career structure for academic GPs and there is a role for the other healthcare partners in primary care training.

Solutions have been proposed for some of these concerns about UK clinical research training[29]. The creation of University Hospital NHS Trusts could provide a new vehicle with responsibility for training doctors, conducting medical research and delivering clinical care. Learning from best practice in integrating basic and clinical research across all disciplines, developing new NHS R&D strands to encompass systematised observation and appraisal, redefining units of assessment in the RAE, ensuring greater transparency in university accounting and tracking procedures so that external funders can clearly see what their investment is buying, should all help to improve the culture of research and training in biomedicine. The recent creation of the Academy of Medical Sciences[33], a concept based in part on the model of the Institute of Medicine in the US also provides an excellent opportunity to bring together the basic and clinical biomedical sciences and the training issues. It is hoped that the initial agenda of the Academy will give priority to genetic R&D and provide impetus to addressing many of the issues for education, training and public understanding that have been raised in this article. The emergence of the new discipline of molecular medicine as part of the long-term strategy towards better health is also of great importance in the international context[34].

It has been predicted[22] that the next 50 years will bring remarkable expansion in human knowledge driven in large measure by advances in genomics and informatics. Harnessing these advances to deliver better health, and ensuring that the pace of progress in biomedical research does not pose serious threat to the future competency of healthcare professionals, represent major challenges for all those concerned with education for the medical profession and the maintenance of an internationally competitive research base whose creativity can be harnessed promptly for the benefit of improved patient care.

Acknowledgement

We thank Sir Keith Peters, University of Cambridge, for helpful discussion.

References

1 Bell J. The new genetics in clinical practice. *BMJ* 1998; **316**: 618–20

2 Genetics Research Advisory Group. *A First Report to the NHS Central Research and Development Committee on the New Genetics.* London: Department of Health, 1995

3 Genetics Research Advisory Group. *The Genetics of Common Diseases. A Second Report to the NHS Central Research and Development Committee on the New Genetics.* London: Department of Health, 1995

4 Welsh Health Planning Forum. *The Impact of the New Genetics on the NHS.* Cardiff: Welsh Health Planning Forum, 1995

5 Medical Research Council. *Research Developments Relevant to NHS Practice, Public Health and Health Departments Policy.* London: MRC, 1995

6 Department of Health. *Our Healthier Nation. A Contract for Health Consultation Paper.* London: Department of Health, 1998

7 Office of Science and Technology. *The Human Genome Mapping Project in the UK. Priorities and Opportunities in Genome Research.* London: HMSO, 1994

8 Fears R, Greig R, Poste G. Human genomic research and innovative medicine. *Sci Parliament* 1994; **51**: 4–5

9 Anderson J, Fears R. *Sustaining the Strength of the UK in Healthcare and Life Sciences R&D: Competition, Cooperation and Cultural Change.* Oxted, Surrey: ScienceBridge, 1997

10 Weatherall D. *The Implications of Genetic Engineering for Medical Practice.* London: Royal Society of Medicine, 1988

11 House of Commons Science and Technology Committee. *Human Genetics: the Science and its Consequences Volume I.* London: HMSO, 1995

12 Rosen R, Mays N. Controlling the introduction of new and emerging medical technologies: can we meet the challenge? *J R Soc Med* 1998; **91**: 3–6

13 Clinical Genetics Committee. *Clinical Genetics Services into the 21st Century.* London: The Royal College of Physicians, 1996

14. Kinmonth AL, Reinhard J, Bobrow M, Pauker S. Implications for clinical services in Britain and the United States. *BMJ* 1998; **316**: 767–70

15 Wertz DC, Fletcher JC, Berg K. *Guidelines on Ethical Issues in Medical Genetics and the Provision of Genetics Services.* Geneva: World Health Organization, 1995

16 Human Genetics Advisory Commission. *First Annual Report.* London: Department of Trade and Industry, 1998

17 Genetic Interest Group. *The Present Organisation of Genetic Services in the United Kingdom.* London: Genetic Interest Group, 1995

18 Welsh Institute for Health and Social Care. *Report of the Citizens' Jury on Genetic Testing for Common Disorders.* Glamorgan: Welsh Institute for Health and Social Care, 1998

19 Brower V. New genetics study-aid for doctors. *Nat Med* 1998; **4**: 379

20 Stephenson J. As discoveries unfold, a new urgency to bring genetic literacy to physicians. *JAMA* 1997; **278**: 1225–6

21 Collins FS. Preparing health professionals for the genetic revolution. *JAMA* 1997; **278**: 1285–6

22 Poste G. Molecular medicine and information-based targeted healthcare. *Nat Biotechnol* 1998; **16** (Suppl): 19–21

23 Marteau TM, Croyle RT. Psychological responses to genetic testing. *BMJ* 1998; **316**: 693–6

24 Doyal L, Gillon R. Medical ethics and law as a core subject in medical education. *BMJ* 1998; **316**: 1623–4

25 Jonsen AR, Durfy SJ, Burke W, Motulsky AG. The advent of the 'unpatients'. *Nat Med* 1996; **2**: 622–4

26 Task Force. *Clinical Academic Careers.* London: Committee of Vice-Chancellors and Principals, 1997

27 House of Lords Select Committee on Science and Technology. *Clinical Academic Careers.* London: The Stationery Office, 1998

28 Beecham L. UK needs an extra 1000 doctors. *BMJ* 1997; **315**: 1487

29 Policy Forum. *Is the Status of Research in UK Clinical Medical Sciences Declining?* London: SmithKline Beecham, 1998

30 National Committee of Inquiry into Higher Education. *Higher Education in the Learning Society.* London: HMSO, 1997

31 Elwyn GJ. Professional and practice development plans for primary care teams. *BMJ* 1998; **316**: 1619–20

32 NHS Executive. *A Guide to Specialist Registrar Training.* London: Department of Health, 1998

33 Dickson D. Britain establishes Academy of Medical Sciences. *Nat Med* 1998, **4**: 375

34 Editorial. Molecular medicine growing pains. *Nat Med* 1998; **4**: 641

Genetics in drug discovery and development: challenge and promise of individualizing treatment in common complex diseases

Klaus Lindpaintner

Roche Genetics, F. Hoffmann-La Roche, Basel, Switzerland

An idea, once born, can never more be undone.

Friedrich Duerrrenmatt, *The Physicists*

About 25 years ago, a fundamental change in the conduct of pharmaceutical research occurred: the baton was passed from the chemist to the biologist as the driver of innovation. Previously, the biologist – who was first and foremost a pharmacologist and physiologist – was relegated to the role of testing new compounds synthesized by chemists. More recently, empowered by rapid progress in cell and molecular biology based understanding of disease, biologists developed the ability to formulate rational hypotheses of disease-relevant mechanisms and began to present the chemist with putative cellular targets that then became the starting point for drug discovery – inverting the previous sequence. Commonly, these targets were derived from a mechanistic physiological understanding of how biochemical pathways affect certain phenotypes. In parallel, the development of tools to determine molecular structure generated great enthusiasm for 'rational drug design'. As the limitations of this approach became disappointingly clear, the emphasis of drug discovery shifted to high-throughput, automated screening of as large and as diverse a collection of molecules as possible. This primary approach, initiated and driven by the biologist, is then followed by a dedicated, systematic effort to improve a 'hit' by the medicinal chemist. Highly potent, specific, and selective drugs thus developed, *e.g.* proton pump inhibitors, HMG-CoA reductase inhibitors, and serotonin re-uptake inhibitors, to name a few, have changed the face of medicine.

Correspondence to: Prof. Klaus Lindpaintner, Roche Genetics, F. Hoffmann-La Roche, CH-4070 Basel, Switzerland

Despite these dramatic strides, a critical dilemma remains: our medicines do not work well for everyone. As physicians in clinical practice will testify on the basis of daily experience, a medicine can be well-tolerated and dramatically effective for Patient A, only to be ineffective or

British Medical Bulletin 1999;**55** (No. 2): 471–491

harmful in Patient B. This cumulative experience of unpredictability with respect to the use of pharmaceutical products shapes physician and patient attitudes enormously, and with good reason. An understanding of this fundamentally unsatisfactory situation also helps to explain the repeated outcry about the high costs of new medicines, which in public discourse greatly overshadows any acclaim for the pharmaceutical industry's dramatic successes. Certainly, in the current climate of cost containment, this 'hit-or-miss' quality of successful prescribing is no longer tenable. The most pressing questions facing the pharmaceutical industry at the end of the twentieth century are deceptively simple ones: what are the reasons that medicines work differently in different individuals, and how can we predict these differences? The answers will be found in pharmacogenetics.

Before grappling with these most fundamental issues, however, one would like to understand the reasons that biology-based efforts to produce effective medicines have led to this state of only partial success. The principal reason is a familiar one in the history of science: the confusion of association with causality. Biology-driven drug discovery is based on observed phenomena that imply and elucidate underlying physiological mechanisms without defining them precisely. The strong association of such a mechanism, or the response to a drug intervention with altered function in an organism does not ultimately provide a causal explanation for the alteration, though it permits the development of better hypotheses. Thus, one can measure improved glucose metabolism after administration of insulin without learning the cause of the insulin deficiency, and without appreciating the existence of factors modifying insulin sensitivity. Similarly, one can administer an antibiotic and observe a bactericidal effect associated with recovery of the patient from pneumonia, without appreciating the role of the immune system in the curative process. When disease-causing mechanisms are simple, this associative approach can suffice. However, when diseases are etiologically heterogeneous certain subgroups of disease-sufferers will respond to a given treatment better than others. In order to understand these differences, scientific techniques that allow a better understanding of causation are essential. Therein lies the promise of strategies that look for explanations at the level of the gene.

With the marriage of long-established principles of inheritance (Mendelian genetics) to newly-found tools of molecular biology, the discipline of molecular genetics was born. A critical shift in the practice and outlook of the biosciences occurred, enabling us to progress closer to a true causative understanding of biology than ever before. By the mid to late 1980s, single gene diseases began to be unravelled at an increasing pace, demonstrating clearly the power of this approach. The initiation of the Human Genome Project shortly thereafter endorsed this concept as one of the critical elements of advancement in the life sciences. The tools thus created have increasingly empowered us to bypass the biochemical intermediaries and to probe the ultimate substrate of biological variance: the genome. In this way, we become increasingly able to understand the differences between healthy and diseased individuals, and also the critical differences among individuals who initially appear to have the same disease. Association at this level becomes, in effect, synonymous with causation.

Will the advances in genetics help us find better ways of detecting, treating, and preventing human disease? Will we be able to individualize the prescribing of medicines in ways that take advantage of genetic variation? We are early in the process of applying many of these new concepts and tools, and much of the data generated so far remains anecdotal, in want of replication and validation. Nevertheless, there exist major opportunities and applications for molecular genetics in the process of drug discovery and development, opportunities that offer a vision of more effective therapies for individuals suffering from our most common diseases. This essay will examine the characteristics of those diseases from a genetic perspective, and describe the short- and longer-term importance of pharmacogenetic strategies in developing improved, individualized treatments.

Mendelian versus complex, common disease

Molecular genetic techniques and approaches have proven most effective in the area of the so-called 'classical' inherited diseases. These are illnesses in which alterations in one gene are of overriding pathogenic importance, and which are transmitted in a simple fashion that follows quite evidently Mendel's laws of inheritance, such as Huntington's chorea, sickle cell anemia, and cystic fibrosis, to name but a few. Thanks to advances in methodology and technology achieved over the past two decades, the concepts of 'reverse genetics' and 'positional cloning' have evolved from pure theory to become standard genetic investigative techniques. Powerful, unbiased 'genome-screening' techniques (which cover the genome with narrowly-spaced, but otherwise randomly assembled collections of markers) were applied to data-bases drawn from genealogically and clinically well-characterized pedigrees of affected families allowing the identification of disease-causing mutations in single genes. Early successes in this approach led to the discovery of the genes underlying the cause of some of the most common single gene disorders, such as cystic fibrosis and muscular dystrophy. These successes dramatically underscored the remarkable power of molecular genetics to find causative disease principles, and established the fact that a quantum leap in biomedicine had occurred[1].

However, in as much as these 'single gene diseases' are devastating for the affected patients and their families, their overall impact on public

Table 1 Heritabilty estimates for some common complex diseases

	Heritability	MZ twin concordance	DZ twin concordance
High blood pressure	0.60	0.6–0.8	0.3–0.5
Asthma	0.72–0.8	0.12–0.89	0–0.5
Type 1 diabetes	0.72	0.35	0.03–0.05
Type 2 diabetes	0.26	0.50	0.37
Type 2 diabetes + IGT	0.61	0.63	0.43
Rheumatoid arthritis	0.32	0.15	0.04

health is in fact extremely small. They pale in importance compared to more common diseases, such as ischemic cardiovascular disease, non-insulin dependent diabetes, osteoporosis, cancer, *etc*., that affect large segments of the population, and that exert the major toll of morbidity, individual suffering, and public health care spending.

We recognize today that these so-called 'common complex diseases' all have both external (environmental) and internal (genetic) contributors. The external contributors are amply documented in the annals of a century of epidemiological investigation: toxic exposures, exercise patterns, dietary habits have all been shown to modify disease risk. The internal contributors or genetic factors have historically been recognized primarily based on the observations of familial clustering of disease and aggregation of cases, in the absence of classic segregation patterns seen in single gene diseases. This clustering became apparent for many of these diseases as a result of epidemiological surveys, and then from more directed twin- and adoption studies (*see* Table 1). For many of these illnesses the establishment of a 'positive family history' has long served as an important parameter for individual disease risk estimation. Thus, for example, stroke occurs twice as commonly in men with a maternal history of stroke than in those without such a history[2]; and hypertension is significantly more commonly shared among natural than among adoptive siblings and parent-child pairs[3]. In contrast to the monogenic disorders mentioned above, these diseases are polygenic: the inheritance patterns are complex, involving the transmission of combinations of disease-susceptibility genes, each of which may contribute in varying degrees to the disease phenotype

Today we appreciate that the manifestation of a common 'complex' disease is the result not simply of an additive accumulation, but of a more complex interplay between several environmental (*e.g.* diet, environmental agent exposure, behavior) and several genetic risk factors that may be categorical (*e.g.* permissive) or quantitative (*e.g.* additive, multiplicative) in nature. Thus, any one of these risk factors alone is commonly not sufficient to lead to the manifestation of disease at a given time during an individual's life. This concept helps to explain the puzzling individual examples of heavy smokers who escape lung disease or alcoholics who remain free of cirrhosis. Genetic factors decrease or increase the risk of experiencing a disease, and thus act as either protective or predisposing. The cumulative effect of many counterbalancing or enhancing influences, both environmental and genetic, ultimately determines overall risk. This will be discussed in more detail below.

The complex pattern of inheritance that these diseases display, *i.e.* one that does not follow the simple rules of an autosomal dominant or recessive transmission, but rather presents itself as accumulation or clustering of disease in families, is at times referred to as non-Mendelian inheritance (a term that is probably more appropriately reserved for epigenetic inheritance). Paradoxically, however, the apparently unpredictable pattern of inheritance in common complex diseases is directly predicted by, and in perfect agreement with, Mendel's second law, describing the independent assortment of alleles; thus, a large number of contributing alleles, each and every one of which follow precisely Mendelian inheritance rules, but assorting randomly, create the familial complex transmission patterns observed. Theoretically, given full knowledge of the effects of all contributing gene variants on a particular phenotype, characterization of an individual's genotype profile should allow a reasonably accurate prediction of his/her disease risk that approaches the reliability possible in monogenic disorders.

However, given the anticipated multiplicity and complexity of genetic factors that, in concert with a similarly complex set of environmental variables, contribute to any common complex disease, it is clear that we will, even under the best of circumstances, be left with a considerable variance around any point-estimate of risk that we arrive at using even the most sophisticated approaches. This is primarily the consequence of the fundamental paradox we are encountering by recognizing disease heterogeneity, and trying to account for it by subdividing larger conventional diagnoses into smaller and smaller subgroups of patients: at the extreme end, this reduces sample size to unity – very much in accordance with the recognized uniqueness of each human being, but with complete lack of the kind of statistical data we traditionally use to test the likelihood that an observation is (biologically) true. Thus, meaningful data will only be obtainable on a limited number of those gene variants that are relatively common and that contribute at least modestly to the phenotype, leaving considerable 'noise' emanating from more rare gene variants, and from those that individually contribute only minimally to the trait (although in sum they may significantly modulate its expression).

An extremely important conclusion follows from these considerations: in the context of complex diseases a 'genetic risk factor' never takes on the kind of deterministic ('Mendelian') quality that the presence of a disease-associated mutation represents for risk assessment in 'classic' monogenic diseases. Monogenic gene mutations can, in first approximation, truly be viewed as disease-causing (although their morbid manifestations are, as is well appreciated, not absolute, with considerable variance often observed among carriers of the same mutant gene within one family — and even more pronounced in different families – with regard to age of onset, severity of symptoms, and clinical course). In contrast, this will never be the case in complex disease. The very fact that we speak of 'disease-susceptibility' and 'disease-predisposing' rather than disease-causing genes bears witness to the far greater modulatory role that un-measured, un-measurable, or un-interpretable (due to missing genetic epidemiological data) factors play in these disorders, as discussed above. Thus, in complex disease, even the most comprehensive knowledge of relevant genetic and environmental factors affecting the likelihood of a disease will inevitably allow us to arrive only at an improved, and certainly more individualized measure of probability, or relative risk, but will never reach the prognostic power typical for simple monogenic disorders. While this improved risk assessment and disease understanding will have major potential impact on patient management and the ability to deliver better and more prevention-oriented care, it will not fundamentally change the nature of diagnostic message — one of relative risk – delivered: thus, the much-feared 'stigmatizing' aspects of a single gene disorder diagnosis will not be encountered for the genetic variants contributing to a common complex disorder. Any one genetic factor will only shift the overall risk, but will never account for all of it, and even the integrative consideration of several genetic factors will not achieve this goal. It should be reassuring to patients that genetic testing in common complex disease will, therefore, not be 'stigmatizing', but will, in essence, deliver information that is qualitatively very similar to disease-risk assessments made today based on a patient's environmental and behavioral profiles; information that, rightly or wrongly, does not encounter major sensitivity in the community. Thus, every smoker understands today that he or she faces a decidedly increased risk for a range of pulmonary diseases, yet, in the community, smoking is not perceived as a sign of impending doom or demise (of note, the relative risk attributable to smoking as a single predisposing factor for these illnesses is almost certainly manifoldly higher than that of even the most important common susceptibility-gene variant!). These facts, however, are not widely appreciated in either the lay or the medical community, and education about these issues is sorely needed.

Since all common complex diseases are of delayed onset, consideration of the time (or age) axis must also be regarded as an essential (permissive or contributing) domain in the consideration of disease causation, and it is indeed likely that risk-lowering interventions will first and foremost delay, but not necessarily altogether avoid, manifestation of an illness.

To understand how these individual susceptibilities to disease can be characterized, a brief discussion of single nucleotide polymorphisms (SNPs) – which are anticipated to be the overwhelming source of such susceptibility-conferring gene variants – will be helpful. Scattered throughout the genome of all species are single nucleotide changes that are the very substrate of human diversity. They occur in humans every 300–2000 base pairs along the genome; in principle, they may occur at any nucleotide, although for genetic epidemiological purposes those that are relatively common will be of greatest interest. The vast majority of these SNPs are functionally silent, occurring in non-coding or non-regulatory regions of the genome. However, some fraction of these SNPs does result in altered protein structure or expression. These (potentially) biologically functional SNPs are considered the essence and substrate of human diversity in both health and disease. Once these common, biologically (measurably) relevant SNPs have been identified, and, in particular after those that contribute materially to disease risk have been characterized, a SNP-based 'genetic profile', which may be viewed as an individual's 'fingerprint' of relative (genetic contribution to) risk for various illnesses will become a feasible proposition. As such, SNPs are of course also to be expected to hold the clue to individual differences in drug efficacy and are a logical place to look for the explanations that we seek.

How will the use of genetic investigation and tools advance our approach towards these common complex diseases, in particular with regard to drug discovery and development? It is likely to have profound impact and to occur in a number of different areas -- and on a number of different levels and timescales. Genetic investigation will affect all stages of the process, from target discovery, target validation and compound selection to clinical development and marketing of a therapeutic agent. We predict that the incorporation of genetic tools and considerations into this process will be one of the pre-eminent forces that will advance the practice of medicine towards a more individualized approach to health care. Ultimately, this will be done through the integration of the processes of diagnosis and treatment: after determining an individual's personal disease profile, a therapy tailored to (the most) relevant individual characteristics can be recommended. The result will be more effective treatments and less dependence on trial and error in prescribing.

Methodological challenges in the understanding of common complex disease

If we contend that understanding the cause of disease should render drug targets that are superior compared with the more purely associative ones we have pursued so far, then we should certainly expect the use of genetic approaches to have a major impact on this level. The obstacles, however, are inherent in the nature of complex, common disease itself. Although the molecular genetics tool-kit for gene discovery has become a commodity, its application to the dissection of these disorders is fraught with major difficulty. The common complex diseases share four characteristic features, each of which contributes its share of methodological challenge. These characteristics are: (i) multifactorial and ecogenetic nature; (ii) polygenic inheritance; (iii) genetic heterogeneity; and (iv) continuous (as opposed to dichotomous) traits.

The multifactorial nature, with external (environmental and behavioral/life-style related) factors being as important, or more so, than the aggregate of the genetic factors, makes careful accounting, or controlling for these variables essential, while at the same time diluting the magnitude of phenotypic variance ascribable to genetics. As described above, the relationship of genetic and environmental factors may be not a simple, additive one, but a complex (ecogenetic) interaction, requiring careful statistical assessment of interaction terms. For example, while excess sodium consumption will modestly elevate almost everyone's blood pressure as a basic, physiological consequence of fluid retention, it can do so much more dramatically in the presence of gene variants that render an individual 'sodium sensitive'. Of course, depending on the relative contribution of nature (genetics) *versus* nurture (environment) -- which varies along a sliding scale from disease to disease – discovery of the genetic component(s) will be more or less difficult (as well as less or more meaningful for disease management); thus, it maybe anticipated that the genetic underpinnings of insulin-dependent diabetes will be more readily accomplished than those of non-insulin dependent diabetes.

Likewise, the polygenic nature of common complex diseases means that disease causation attributable to genetic contributors represents the aggregate effect of several genes, making the effect for most single contributing genes rather modest and difficult to detect. In addition, consideration must be given to the interplay of genetic factors that may be more complex (epistasis) than is captured by linear modeling.

Genetic heterogeneity means that multiple genes in different combinations may contribute to an apparently identical clinical presentation. For example, there might be a dozen gene variants that contribute to osteoporosis. While one patient may carry the first eight, but not the remaining four, the next patient may carry the last six, thus sharing only 2 genetic risk factors with the first one. Clearly, this genetic heterogeneity of what appears to be a single disease entity, adds statistical artifact commensurate with the degree to which the set of genetic factors involved in disease causation differs from one person or one family to the next.

Lastly, it is important to bear in mind that common complex diseases are characterized by quantitative, continuous traits that do not offer the luxury of simple dichotomous categorization. The reason for this becomes clear when we consider the additive (or more complex interactive) influences -- either in the direction of disease predisposition or prevention – that a set of genes and their respective variants may have on a clinical phenotype. Let us assume that there is a dozen gene variants that tend to raise blood pressure (BP) and another dozen that tend to lower BP. Person A might have variants that increase the activity of 8 BP-raising genes and only 3 genes that lower BP, resulting in elevated blood pressure compared with the population average. Person B might have 5 BP raising genes and 5 BP lowering genes, and map to the apex of the Gaussian curve. And person C might carry 2 BP-raising and 6 BP-lowering gene variants and be assigned to the lower tail of the distribution. Using this example, it becomes already clear that there are, strictly speaking, no self-evident criteria for any of the common complex diseases to declare 'normal' or 'abnormal', 'diseased' or 'healthy' status, and when we use these categorizations (as we pragmatically must) in the practice of medicine, they are based on more or less arbitrary definitions and cut-off values in comparison with large samples. By binning quantitative data into categorical, a great deal of information is lost, and much potential error introduced. For example, a measurement error of 2 mmHg on a poorly calibrated sphygmomanometer that reads 141 mmHg instead of 139 mmHg can categorize a patient falsely as hypertensive and thus radically change his or her clinical status from control to case, when in a quantitative analysis the effect of this error would have been hardly noticeable. Limiting case or control status, if one must carry out a study based on dichotomous variables, to more extreme tails of the distribution is somewhat helpful, but presupposes the availability of originally quantitative data and also ignores a large part of the spectrum of clinical variance. Similar examples as drawn here from blood pressure/hypertension abound, *e.g.* for serum glucose or cholesterol, for atherosclerotic changes in vessels, or for neurofibrillary tangles in aging brains. In evaluating the importance of these conditions, painstakingly precise measurements are essential to capturing the relevant phenotypical information, and analyses based on sloppy measurements are obviously worse than useless.

Considered together, these four considerations help explain why these disease entities are called 'complex'. To the biometrician, the elements enumerated above amount to a statistical nightmare, where modest signals are pitched against a multitude of sources of noise. How might one tackle this problem, in order to identify new potential targets for drug development?

Long-term impact: new target discovery

One way of approaching this dilemma is to use reductionist approaches, tried and tested by experimentalists in all branches of science to reduce noise, at the potential cost of failing to capture the truly relevant components of a phenomenon. The use of inbred animal models of disease represents such an approach — it allows an investigator to reduce both environmental and genetic heterogeneity, the former by tightly controlling the environment, the latter by working with inbred disease strains in which all individual animals share exactly the same set of disease-contributing genes. Most importantly, of course, programmed breeding protocols can be implemented, allowing the creation of very large and powerful 'pedigrees' that are impossible to find in any human population. The price of this methodological reductionism is the uncertainty of whether any extrapolation from the experimental to the human condition is justified. Still, novel disease mechanisms and pathways thus identified may provide valid new targets for drug discovery. Among human populations, a similar scenario of reduced complexity can be found among so-called founder populations, or genetic isolates, where an expanding population can be traced back to a more or less recent, small original ancestry, thus limiting the degree of genetic diversity (*i.e.* heterogeneity for a given disease or trait) present in the sample. Recent efforts to find asthma-related genes among the settlers of Tristan da Cunha, a remote island in the South Atlantic[4], as well as long-standing genetic investigations in the Finnish-Karelian, the Icelandic, and the Amish populations[5], are prominent examples of this approach.

An alternative approach towards dealing with very complex data sets lies, conceptually, in collecting massive amounts of data that provide sufficient statistical power to cancel out noise and to reveal the signal. Large genome screens carried out in extensive collections of nuclear families in the search for common complex disease genes represent an early, and still relatively modest, example of such an approach. Presently, available off-the-shelf technology already allows significant scale-up of such approaches. Limiting factors to this approach are processing cost per sample and, much more significantly, the difficulty of identifying and ascertaining sufficiently large proband cohorts. Such 'brute force' approaches will also require enormous capacity for data handling and analysis. Novel biomathematical approaches that are capable of handling very large and complex datasets, such as clustering or principal component analytical methods, will play a major role in our ability to interpret the data generated and transform it into meaningful information. This is particularly true for the envisioned application of very large sets (10s to 100s of thousands) of random genetic markers (rSNPs) in large case-control cohorts[6] for genome-wide association (linkage disequilibrium) studies; for this to become reality, however, SNP processing technology will need to advance by orders of magnitude upwards (to something like 10^6–10^8 genotypes per day), paralleled by an equally significant decrease in cost, to fractions of a cent per genotype.

While intelligent arguments can be made in support of the primacy of 'causal' targets over those that are only associative or simply deduced from function, it should be also be emphasized that this does not by necessity imply that only individuals in whom a particular causative disease mechanism is operative or predominant will respond to a medical intervention aimed at that mechanism. Rather, by finding a gene (and, in due course, the associated additional elements of the biological pathway or signaling cascade) that, if dysregulated, has the potential of inducing the morbid phenotype of interest (at least in a subgroup of patients with the disease), one generates important new functional and mechanistic-pathogenetic knowledge about the disease. Based on our current experience, there is a very good likelihood that a drug targeting such a mechanism -- if it normally participates in the regulation or homeostasis of a disease-relevant phenotype or trait – will be at least partially effective also among individuals in whom a different pathomechanism underlies the same disease, thus broadening the pharmacological arsenal available to any patient suffering from this disorder.

The use of causative targets will not necessarily accelerate, in an individual case, the time needed for discovery and development of a new drug, since most of the steps involved in proceeding from target to marketed medicine remain unchanged. Indeed, we may anticipate that a gene or its encoded protein, even if identified as 'disease-contributing', may not necessarily be a tractable drug target at all, and that other elements or members of the pathway will first need to be elucidated. However, it is safe to assume that the drug discovery process, as a whole, will become more efficient as putative drug targets are identified by genetic approaches. These targets and/or associated pathways come with the intrinsic validation of being disease-relevant. They may provide enhanced knowledge about the structural effects of genetic polymorphisms and thus can guide compound selection. For these reasons, the rate of failures and 'killed' projects is likely to decrease as more such targets are used. Overall, therefore, the drug discovery process will become more expeditious, more efficient, and thus more economical.

Mid-term impact: target validation and compound selection

On a more intermediate-range time scale, the deployment of molecular genetic approaches has the potential of significantly contributing to the drug discovery process by improving the likelihood of selecting compounds that will ultimately prove successful and by providing guidance for the clinical development process. In addition, there is certainly value in validating the target as disease-relevant, by showing an association between a gene variant and disease prevalence or incidence, even at the relatively late stage of clinical development.

An important mid-term goal is clearly the screening of active and contemplated drug targets for the presence of genetic variants. If genetic variants are found, testing for potential impact of such variants on disease, disease susceptibility, and on chemical tractability will be the logical next steps. Screening the targets for SNPs is thus widely considered a useful, perhaps essential strategy. Polymorphisms found in the target, particularly if they affect the translated amino acid sequence, should be assessed with regard to their possible impact on target structural confirmation, and its possible effects on target-ligand binding/interaction. In addition, even if they are *a priori* 'silent', they should be followed-up with genetic epidemiology studies to compare the prevalence of the respective variants in cases (patients with the disease in question) and controls (since they may be in linkage disequilibrium with non-captured variants affecting gene regulation, or they may directly affect mRNA stability and or translational efficacy).

If such an association between a molecular variant of the target and the disease is indeed found, at a sufficient, predetermined level of statistical significance and/or magnitude of effect, this information may have important consequences both with regard to the chemical screening strategy as well as to the design of clinical studies. Presumably, a compound that selectively targets the disease-associated molecular variant of the target could be designed and tested. The above-described strategy would certainly validate the drug target as a disease relevant one, and presumably would increase the likelihood of successful completion of the development efforts. Thus, a small molecule with high, selective affinity (if the mutation affects binding) to the target variant of interest, but not to the variant found in non-affected probands, might be pursued, and a clinical development strategy may be considered that preferentially or exclusively enrolls carriers of the disease-associated target.

In contrast, if no association between the target's genetic variant(s) and disease prevalence is found, and assuming that there is at least a possibility that the mutation affects drug binding, it might be wise to conduct secondary screens with assays employing the molecular variant(s) of the target to ensure equally optimized pharmacodynamic characteristics for all variants and to avoid unwelcome surprises later on during clinical development in the form of pharmacogenetic phenomena that increase response variance, thereby decreasing study statistical

power, and in turn increasing development cost, and potentially fragmenting the future market of the drug.

Carried out in a more comprehensive fashion, this 'frontloading' of target polymorphisms may also include other members of the targeted pathway, as well as other 'candidate' genes considered to be potentially interacting with the mechanism of action of the drug, or its metabolism.

Short-term impact: pharmacogenetics

'Idiosyncratic' individual differences in drug response have long been recognized as a major limitation of successful and safe medical treatment. In the absence of identifiable factors that would allow the prediction of a specific response these 'differences' have been a source of major frustration for patients, for physicians, for third party payers, and for the pharmaceutical industry. Given the, so-far, only possible trial-and-error approach to finding the appropriate drug exposes the patient to delays in receiving appropriate treatment for his or her condition, increases the workload on the physician, and represents wastefully spent resources on ineffective treatments for the health care payer. The industry sees large investments wasted because of adverse events that cannot rationally be avoided, and because the dilutional effect of non-responders may prevent a clinical trial from reaching the statistical significance needed for regulatory approval. The field of pharmacogenetics seeks to understand the genetically encoded inter-individual differences among patients that are at least one important source of these differential responses. A better understanding of the nature of these differences, and the ability to test for them, are, therefore, increasingly being viewed as not only sensible, but as a prerequisite for selecting the appropriate drug for the patient on an individual level.

Pharmacogenetic phenomena can conveniently be separated into two principal groups, those that affect the metabolism of a compound (pharmacokinetics) and those that affect the activity, and thus the efficacy of a medicine (pharmacodynamics). The former have of course been recognized for many years, based on biochemical observations and assays, and are considered routinely during the development of a drug. Thus, phase I trials are commonly carried out in appropriate, well-defined cohorts with representation of specific metabolic phenotypes, *e.g.* fast and slow acetylators. The recognition of a substantial number of variants of enzymes that play important roles in drug metabolism based on DNA polymorphisms opens the way to a more comprehensive, more precise, and more sensitive examination of these interactions than has been possible on the level of protein biochemistry. Understanding the interaction between certain metabolizing enzyme variants provides not only the opportunity to avoid adverse events, but also to recapture efficacy in relevant patients by adjusting dosage and/or choosing optimized formulations.

The recognition of genetic variants' role in affecting drug response or efficacy is of more recent date, and while conceptually attractive and clearly to be expected, remains at this point mainly anecdotal and in need of replication and validation. For example, the acetyl cholinesterase inhibitor, tacrine, has been reported to show differential efficacy dependent on ApoE genotype (as well as a superimposed sexual dimorphism) in patients with Alzheimer's disease[7,8], and pravastatin has been reported to provide differential benefit to patients with coronary heart disease stratified by cholesteryl-ester transfer protein (CETP) genotype[9]. Given the complexity of the disorders for which such examples have been reported, the clear-cut, categorical results shown seem surprising, and additional data will certainly be required. However, it is of some interest to speculate that, in the presence of very powerful environmental stressors, which are orders of magnitude stronger than those encountered in the natural environment, the effect of certain genetic factors may be greatly enhanced, and thus become much better discernible. Viewed from this angle, pharmacological agents could certainly be seen as such 'non-physiological stressors'. If this concept were indeed to emerge as a common principle, then we may ultimately be confronted with significant pharmacogenetic effects far more often than would be expected based on differential, modest contribution of individual gene variants to complex disease entities.

Many pharmaceutical companies are today seriously considering, or already engaged in, the collection of a DNA sample from each of their patients participating in phase I, II, or III trials, to retain the option of investigating potential associations between genetic variants and positive (efficicay), negative (lack thereof), or adverse clinical responses. Currently, these approaches are restricted to a limited number of likely or suspected 'candidate genes' (generally, these are disease-related and/or drug-specific). The ongoing efforts to develop a comprehensive, densely spaced, genome-wide SNP map may allow us in the future to conduct screens for pharmacogenetically active genes as whole-genome, unbiased searches.

What are the consequences of discovering a pharmacogenetic interaction? On a conceptual level, we may be able to avoid exposing side-effect-prone individuals to the drug, or, in the case of a pharmacokinetic phenomenon, to adjust the dose accordingly. Regulatory authorities are beginning to recognize this and have issued early language encouraging pharmaceutical companies to develop rational risk-stratification approaches based on pharmacogenetic testing. This information may allow them to receive regulatory approval that would otherwise not be granted. Likewise, we may avoid prescribing a drug to patients who are unlikely to respond, restricting it to a genetically defined group of responders. Defining such a subpopulation of responders might thus allow the 'rescue' of a drug that would – in a non-stratified population – not cross the statistical threshold of efficacy required for regulatory approval.

Commonly, concerns are raised that such a scenario may not be economically viable: by 'fragmenting' indications into smaller (sub)-segments, the critical mass of revenue that justifies and enables an expensive drug development program may not be reached. Enrolling patients stratified by genotype will slow down recruitment, and the need to provide a diagnostic along with the pharmaceutical will further complicate the logistics of marketing and distribution. However, it is extremely difficult to sustain the argument that a responsible company should market products to individuals for whom they will be harmful or ineffective, if the means exists to predict these results. Furthermore, on a practical level, the effort to produce medicines with superior therapeutic efficacy and fewer side effects will likely be repaid by improved patient adherence to therapy (compliance), by higher market segment penetration, and by longer lasting treatment. Moreover, if the genotype-specific response is one tied to genotype-specific disease etiology, then one may speculate that predisposition testing for this genotype may extend the size of the target audience to include less symptomatic, earlier cases, and perhaps even asymptomatic individuals (prevention). It stands to reason that a substantially more effective drug would also command a premium price; under those circumstances, the companion diagnostic could, of course, also generate additional revenues. In addition, if a specific responder subgroup is known and targeted, then restricting enrollment to this group should allow significant savings by allowing much smaller trials to be carried out, provided that safety requirements are met. Consideration of pharmacogenetics and the performance of appropriate studies is likely to become very quickly as much standard part of a drug development process as detailed pharmacokinetic and drug–drug interaction evaluations have become by now; and it is not unlikely that regulatory authorities will one day require such data to be submitted. Indeed, in as much as it should be in every company's interest to identify important pharmacogenetics phenomena as early as possible to retain the opportunity to adjust regulatory submission and marketing strategies in time, there is utility even in the two extreme outcomes of this policy: if sufficiently complex pharmacogenetics effects are noted to make the marketing of a drug not feasible, killing the project at the earliest possible timepoint is the only logical and money-saving consequence; and demonstrating complete absence of any measurable pharmacogenetics effects will serve as a strong argument for maximally broad marketing of the drug.

The current emphasis on cost-effectiveness for new therapies will dictate that every effort be made to produce medicines which are as useful as possible for their targeted recipients. Thus, pharmacogenetics will find itself positioned at the very core of the development of new medicines, and companies that fail to recognize this will have to learn the hard way.

Towards individualized, integrated health care: medical, ethical, economic implications

In the broadest of terms, the consequences of using genetically driven approaches towards target discovery and drug development will result in an increasingly 'individualized' approach towards health care, logically so because genetics is, after all, the door through which we gain access to inter-individual differences. As we are learning to consider these differences as important elements in the approach to the patient, and in our ability to deliver effective and successful health care, we will invariably arrive at a more and more 'custom-tailored' practice of medicine. Conceptually, this is nothing new: ever since the days when blood-letting was practiced as near-universal therapy, an increasingly differentiated and sophisticated approach towards classifying separate clinical entities, *i.e.* differential diagnosis, has been the driving force behind all progress of medicine. The incorporation of genetic data simply adds yet another level of sensitivity and specificity to the diagnostic armamentarium that is in everyday use by the practicing physician today, albeit with certain special features. Inherent in this development is an increasingly important role of diagnostic tests in the practice of medicine: as therapy tends to target more and more specific subsets of disease entities, and of the overall patient population, increasingly precise diagnostic tests that help discern these subcategories are the obvious prerequisite. If the prescription of a drug requires knowledge of a certain biochemical or genetic marker, then only a fully integrated 'package deal' will be viable in the market, providing the diagnostic, the therapeutic, and --importantly – the specialized information needed for patients, physicians, and other health care personnel to be able to use and implement such a sophisticated, highly personalized approach.

Genetic information remains a breed apart from all other kinds of medical data, with regard to the trepidation with which the general public views it. This reserve, which has its roots in the historical association of eugenic movements with the academic discipline of genetics, as well as in the fear that genetic data will be misused by insurance companies, employers, *etc.* to discriminate against individuals 'branded' as being at risk for certain diseases, represents a major stumbling block for both genetic research and the use of genetic data in the clinical setting. Requirements for confidentiality and limitation of using the data, even in the research setting, only in the precise context specified by the probands, place major constraints on our ability to generate meaningful data. While genetic data differ from other health care data in their prospective and prognostic information content, and in the potential implications their knowledge or disclosure may have for blood relatives of the tested probands, these characteristics have real meaning only in the case of 'classical' single gene disorders. In contrast, the role of genetic factors in common complex disease, as we have seen, is a much less discrete one, providing, therefore, never more than probability estimates of associated disease risk. Insurers have, for many decades, been asking applicants about their family medical history, and have thus had access to an integrated parameter of heritable risk whose reliability far exceeds that of any of the presently available genetic tests in complex disease. This practice never led to an outcry of protest. Clearly the need for education, information, and teaching about genetics is large and evident, to put the risks in perspective and illuminate the potential benefits of our growing knowledge.

An aspect that currently occupies many genetic epidemiologists' thinking should not go unmentioned: great emphasis is frequently placed on the importance of measuring SNP allelic frequencies among different 'ethnic groups' and 'races'. Although there is clear evidence for differential allele frequencies among individuals from different ethnic backgrounds (a consideration that used to be considered important in forensic testing until sufficiently complex panels of markers had been designed), a focus on these differences harbors the great danger of re-defining race on the basis of genetics, and not, as we have come to appreciate, on the sociocultural and socio-economic level (in as much as the increasing intermixing of populations from different backgrounds should make a biological definition of race increasingly meaningless, anyway). Indeed, genetics may be viewed as a unique opportunity to move (certainly in the field of biomedicine) away from ethnic/racial stereotypes: understanding the genetic variant underlying sickle cell disease has allowed us to use that parameter, rather than skin color, as a (much more accurate) predictor of disease risk; along the same lines, antihypertensive treatment targeting sodium-sensitive disease will be applied to carriers of the relevant gene variant, once found, based on a genetic diagnosis irrespective of 'race', and no longer, as is current practice – for lack of molecular-genetic definition of the trait – in a blanket, first-line approach to hypertensive

African-Americans. Thus, there is great cause for optimism that the appreciation of the much more profound genetic diversity that is bound to be present at the inter-individual, as compared to the inter-ethnic level will relegate the use of ethnic stereotyping to the closet. This will, however, require a very detailed knowledge of disease contributing genes; the reason that at least in gene discovery and pharmacogenetics trials ethnic stratification is currently still to some extent defendable relates to our inability of accounting on a genotype level for 'genetic background' and/or 'modifier genes', with ethnicity (if representative of a relatively old and homogeneous ethnic group) serving as a integrative parameter for such differences. The perceived need for characterizing random mapping (in contrast to biologically relevant) SNPs for allele frequencies among different ethnicities is more difficult to understand, as we already know that only about 20% of such SNPs show significant differences across ethnic groups; given the large number of SNPs currently being generated, these could simply be ignored.

Although questions surrounding data confidentiality and protection of patient privacy tend to dominate current discussions about ethical and societal aspects of genetic testing, it is important to consider that, in order to derive benefits from genetic information, complete confidentiality cannot and must not be the goal. Prescription of any drug that shows important pharmacogenetic interactions requiring genetic testing will immediately disclose, by inference, the recipient's genotype to the pharmacist and the insurer. The presumably confidential diagnosis of diabetes mellitus is revealed in a similar way in current practice by the filling of a prescription for insulin. Arguably (or maybe not even?), it would make much more sense to disclose the genetic risk/predisposition, once genetic profiling is possible, to the pharmacist and insurer by submitting a prescription for a drug that will prevent the onset of the disease. Rather than treating genetic information as a special form of medical information, standards of confidentiality and individual choice regarding the use of all types of medical data need to be established and/or redefined. Ultimately, society will benefit most by simultaneously acting to protect not so much the privacy of medical information, but the way it is used, thus paving the way for patients to derive the benefits of the progress made in pharmacogenetics and other areas of medicine. This will be best accomplished by enacting appropriate laws that regulate the use of genetic, as well as of all other potentially sensitive medical information, and to make any use of this information outside the terms thus established unlawful and prosecutable. Lawmakers and ethicists in many countries are recognizing the need for such a legal framework and a multitude of efforts are currently underway to draft such legislation; coordinating these activities on an international level would be of great importance to avoid a Babylon of non-aligned rules that will make the use of genetic and other medical data both for research and for medical practice very difficult.

Whereas the prospective or predictive dimension of genetic testing brings a number of ethical and legal considerations more sharply into focus than other diagnostic modalities, it also offers the opportunity to shift medical intervention from therapy to prevention, and from diagnosis to prediction. As we develop a more comprehensive understanding of the contribution that genetic risk factors confer to the overall risk of experiencing a wide variety of health problems, we will be able to offer a custom-tailored program geared at a continuity of health maintenance. Thus, at the outset, predisposition screening would profile an individual's genetic risk for a range of diseases. In many instances, preventive counselling will already be possible and indicated at this stage, with an emphasis on risk reduction by environmental and behavioral modification. In a follow-up phase, the subject could be monitored pragmatically in a targeted, focused fashion with appropriately graded scrutiny -- directly proportional to the determined risk for disease, avoiding the kind of much more dogmatic approach used today that, inspite of all its benefits, is fraught with inefficacy. This strategy will result in a more cost-effective health maintenance program with a better chance of identifying disease at earlier clinical stages where outcomes are likely to be better. Thus, viewed in this fashion, the use of genetic testing and information is supported and justified on the level of both macro- and micro-ethics.

A repeated caveat is necessary lest the reader incorrectly believes that genetic information yields the perfect ability to predict disease. For complex disease, unlike for single gene disorders, an understanding of genetic contributors to disease-susceptibility enables us to be better prognosticators. However, disease risk still will be expressed in terms of statistical probabilities, meaning that small numbers of patients designated at low risk will experience disease while some patients will still fail to be identified. However, the added forecasting precision that is achieved by effectively shifting the basis for preventive strategies from the large-scale epidemiological (population) to the much smaller-scale genetic (individual) level will, if applied comprehensively and consistently, provide a medically and economically brighter future of health care for all participants.

Outlook

A bold picture of emerging paradigm shifts in medicine, pharmacology, and the role of the pharmaceutical industry has been presented. How likely will it become a reality within our lifetime? Let us not forget that while the technical basis for these developments is in place today, and while conceptually the changes outlined appear logical, two major hurdles will need to be overcome: the first is the creation of the knowledge base that is required to carry out profiling for genetic risk — an enormous and dauntingly difficult task – and the second relates to acceptance of these new approaches among the general public and the patients. Both issues are indeed likely to be closely linked: whereas an increasingly powerful database will provide clear evidence of conferring benefits sooner, this will support acceptance of the idea of genetic testing among patients. The experience with patient advocacy groups for single gene disorders shows that efforts to find the causative gene usually find strong support, being recognized as the first, essential step towards treatment, cure, or prevention. We may ultimately witness patients emerging as the driving force behind the development of this vision of integrated and individualized medicine, in an almost radical shift from their currently displayed skepticism, since they clearly have so much to gain.

If we are to be successful in making the use of genetic information a valuable and useful tool, we need to embark on this quest as a coordinated and interdisciplinary effort at a large scale. Only a functional collaboration among clinicians, geneticists, epidemiologists and biomathematicians, as well as lawmakers, bioethicists, sociologists, clergy, teachers, and patient advocates will ultimately succeed in creating the knowledge base and popular understanding needed for progress in medicine to take place. While the tasks in the field of science loom large, those in the field of societal dialogue are no smaller challenge. However, the very significant benefits both to patients and to society that individualized medicine will bring make it critically important for all concerned to work together to allow the growing understanding of the molecular basis of disease to help patients around the world.

Acknowledgements

I wish to thank Drs Lyn Singer-Lindpaintner, Jonathan Knowles and Christina Dahlstroem for their contribution in discussing and editing the contents of this manuscript.

References

1 Collins FS. Positional cloning: let's not call it reverse anymore. *Nat Genet*, 1992; **1**: 3–6

2 Welin-L, Svardsudd-K, Wilhelmsen-L, Larsson-B, Tibblin-G. Analysis of risk factors for stroke in a cohort of men born in 1913. *N Engl J Med* 1987; **317**: 521–6

3 Biron P, Mongeau JG, Bertrand D. Familial aggregation of blood pressure in 558 adopted children. *Can Med Assoc J* 1976; **115**: 773–4
4 Slutsky AS, Zamel N. Genetics of asthma: the University of Toronto Program. *Am J Respir Crit Care Med* 1997; **156**: S130–2
5 Morgan K, Holmes TM, Schlaut J *et al.* Genetic variability of HLA in the Dariusleut Hutterites. A comparative genetic analysis of the Hutterities, the Amish, and other selected Caucasian populations. *Am J Hum Genet* 1980; **32**: 246–57
6 Masood E. As consortium plans free SNP map of human genome [news]. *Nature* 1999; **398**: 545–6
7 Poirier J, Delisle CM, Quirion R *et al.* Apolipoprotein E4 allele as a predictor of cholinergic deficits and treatment outcome in Alzheimer disease. *Proc Natl Acad Sci USA* 1995; **92**: 12260–4,
8. Farlow MR, Lahiri DK, Poirier J, Davignon J, Schneider L, Hui SL. Treatment outcome of tacrine therapy depends on apolipoprotein genotype and gender of the subjects with Alzheimer's disease. *Neurology* 1998; **50**: 669–77
9 Kuivenhoven JA, Jukema JW, Zwinderman AH *et al.* The role of a common variant of the cholesteryl ester transfer protein gene in the progression of coronary atherosclerosis. The Regression Growth Evaluation Statin Study Group. *N Engl J Med* 1998; **338**; 86–93

Index

Micronutrients in health and disease

Scientific Editors

Ian Bremner, Peter J Aggett and Philip James